Global Navigation Satellite Systems

Global Navigation Satellite Systems

Sachin Chandran
Veena V S
Thomaskutty Mathew

2026
Agri Horti Press
New Delhi-110 084

ISBN: 978-93-49690-21-9 (Hardbound)

Published By

Agri Horti Press

New Delhi

E-mail: agrihortipress@gmail.com

Digitally Printed at : Replika Press Pvt. Ltd.

Preface

Global Navigation Satellite Systems (GNSS) is written to provide a comprehensive introduction to the principles, technologies, and applications of satellite-based positioning systems that have become indispensable to modern life. From guiding aircraft and ships to enabling everyday smartphone navigation and precision agriculture, GNSS has revolutionized the way we perceive and interact with geographic space.

This book is intended for students, researchers, engineers, and professionals working in the fields of geospatial science, surveying, civil engineering, telecommunications, transportation, and related disciplines. It aims to build a strong conceptual foundation while also offering practical insights into the functioning and use of GNSS technologies.

The content covers the major global systems—GPS (USA), GLONASS (Russia), Galileo (EU), BeiDou (China)—as well as regional augmentation systems and their interoperability. We begin with the historical evolution and architectural components of satellite systems, including satellite orbits, signal structures, and ground control segments. Core concepts such as positioning, timing, signal errors, and correction methods (e.g., differential GNSS and real-time kinematics) are explained in depth.

Beyond the technical foundations, the book explores the diverse and rapidly growing range of GNSS applications across sectors such as transportation, environmental monitoring, disaster management, geodesy, and autonomous systems. Emerging trends in satellite navigation, including integration with inertial sensors, Internet of Things (IoT), and artificial intelligence, are also addressed.

Special emphasis is placed on error sources, accuracy enhancements, system design considerations, and future challenges in ensuring global coverage, resilience, and security. Case studies, diagrams, and practical examples are included throughout to enrich the learning experience and illustrate real-world relevance.

We hope that Global Navigation Satellite Systems will serve as a valuable resource for learners and practitioners alike, fostering a deeper understanding of the complex but fascinating world of satellite navigation, and inspiring innovative uses of this powerful technology.

Sachin Chandran

Veena V S

Thomaskutty Mathew

Contents

Section 1
Fundamentals of GNSS

Chapter-1

Relativity Laws for the Variation of Rates of Clocks Moving in Free Space and GPS Positioning Errors Caused by Space-Weather Events

1. Introduction

Relativity theory plays a key role in the operation of the Global Positioning System (GPS). It relies on Einstein's second postulate of the special theory of relativity (STR) [1] that states that the speed of light in free space is a constant (c), independent of the state of motion of both the source and the observer. As a result, it becomes possible to measure distances using atomic clocks. Indeed, the modern-day definition of the meter is the distance travelled by a light pulse in c^{-1} s [2]. The value of the speed of light in free space is simply defined to have a fixed value of $c=2.99792458 \times 10^8$ ms^{-1}. As a consequence, the distance Δr between two fixed points is therefore obtained by measuring the elapsed time Δt for light to traverse it and multiplying this value with c, i.e., $\Delta r = c\ \Delta t$.

A key objective in the GPS methodology is to accurately measure the distance between an orbiting satellite and a given object. This type of basic information for a series of satellites can then be used to accurately determine the location of the object [3]. The measurement of the elapsed time for light to pass between two points requires communication between atomic clocks located at each position. The problem that needs to be overcome in the GPS methodology is that the rates of the clocks carried onboard satellites are not the same as those that are stationary on the Earth's surface. For this reason it is imperative that we have a quantitative understanding of how the rates of clocks vary with their state of motion and position in a gravitational field. This topic will be covered in the next section of this chapter.

It also must be recognized that the speed of light only remains constant as it moves through free space and also remains at the same gravitational potential. This is an important distinction that has significant practical consequences when the light pulses must pass through regions in which space-weather events are occurring.

Experimental and theoretical studies of the state of the ionosphere and physicochemical processes that occur in it are caused by the necessity of reliable functioning of communication channels in different frequency bands. In recent years, most attention is paid to the improvement of satellite communication and navigation systems that use transionospheric communication channels. The reliability of communication and navigation systems using ionospheric communication channels depends mainly on the knowledge of the iono- sphere behavior both in quiet and perturbed conditions. This determines the situation when not only ionospheric plasma perturbations related to the dynamics of the atmosphere but also processes related to electromagnetic wave interaction with neutral atoms and molecules of the medium in which the wave propagates should be treated as inhomogene- ities. On the other hand, an analysis of disturbances and failures of operation of space communication systems that use the decimeter wave range and the development of theoretical concepts of physical processes responsible for these phenomena gives new information about the state of the medium and provides the opportunities for further improvement of communication systems. This led to the necessity of the development of special experimental techniques for studying the ionosphere in order to determine the physical reasons for GPS signal delays. In the second part of this Chapter the role of Rydberg atoms and molecules of neutral ionospheric plasma components excited in collisions with electrons in the formation of UHF and infrared radiations of the E – and D– layers of Earth's ionosphere is discussed. A new physical mechanism of satellite signal delay due to a cascade of re-emissions on Rydberg states in the decimeter range is suggested.

2. Time dilation and length contraction

One of the key results of Maxwell's theory of electromagnetism introduced in 1864 was the finding that the speed of light waves has a constant value that is given by the relation: $c=(\varepsilon_0\mu_0)-0.5$, where ε_0 is the electric permittivity and μ_0 is the magnetic permeability in free space. This result appeared quite strange to physicists of the time because it immediately raised questions about how the speed of anything could be the same in all inertial systems. It clashed with the idea that there was an aether in which light waves are permanently at rest.

Maxwell's equations are not invariant to the Galilean transformation and therefore seemed to be inconsistent with the relativity principle (RP). Voigt [4] was the first [5] to give a different space-time transformation that did accomplish this objective. It differed by a constant factor in all of its equations from what is now known as the Lorentz transformation (LT). In 1899 Lorentz wrote down a general version of this transformation [6]:

$$\begin{aligned} t' &= yc\,(\text{t-}vxc^{-2}) && a \\ x' &= yc\,(x\text{-}vt) && b \\ y' &= cy && c \\ z' &= cz && d \end{aligned} \qquad (1)$$

The space-time coordinates (x, y, z, t) and (x', y', z', t') for the same object measured by observers who are respectively at rest in two inertial systems S and are compared in these equations. It is assumed that is moving along the positive x axis relative to S with constant speed v, and that their coordinate systems are coincident for $t = t' = 0$ $[\gamma = (1 - v^2c^{-2})^{-0.5}]$. Voigt's value for ε is y^{-1}. However, Lorentz pointed out than any value for ε in the general

Lorentz transformation leaves Maxwell's equations invariant and thus there is a degree of freedom that needs to be eliminated on some other basis before a specific version can be unambiguously determined. This degree of freedom will be important in the ensuing discussion of relativity theory given below.

The form of the above equations suggested that the Newtonian concept of absolute time needed to be altered in order to be consistent with Maxwell's electromagnetic theory and corresponding experimental observations. Unlike the simple $t = t'$ relation assumed for the Galilean transformation, it would appear from eq. (1a) that the space and time coordinates are mixed in a fundamental way. This meant that observers in relative motion to one another might not generally agree on the time a given event occurred. Poincaré [7] argued in 1898, for example, that the lengths of time intervals might be different for the two observers, i.e. $\Delta t \neq \Delta t'$, and therefore that two events that occur simultaneously in S might not occur at the same time based on clocks that are at rest in. This eventuality has since come to be known as remote non- simultaneity. The space-time mixing concept also predicts that the rates of moving clocks are slower than those of identical counterparts employed by the observer in his own rest frame (time dilation). Larmor [8] published a different version of eqs. (1a-d) (with $\varepsilon = 1$) in 1900, as well as a proof that one obtains the same prediction for relativistic length variations (hereafter referred to as FitzGerald-Lorentz length contraction or FLC) on this basis that both FitzGerald [9] and Lorentz [10] had derived independently in 1889 and 1892, respectively, to second order in vc−1 [5]. Time dilation and remote non-simultaneity also follow directly from Larmor's version of the Lorentz transformation.

In 1905 Einstein's paper [1] on what he later referred to as the special theory of relativity (STR) became the focus of attention, although at least a decade passed before his ideas gained wide acceptance among his fellow physicists. He came out strongly against the necessity of there being an aether which serves as the medium in which light is carried. He argued instead that such a concept is superfluous, and likewise that there is no space in absolute rest and no "velocity vector associated with a point of empty space in which electromagnetic processes take place" [11]. He formulated his version of electromagnetic theory on the basis of two well- defined postulates, the RP and the constancy of the speed of light in free space independent of the state of motion of the detector/observer or the source of the light. Starting from this basis, Einstein went on to derive a version of the relativistic space-time transformation that leaves Maxwell's equations invariant. It is exactly the same transformation that Larmor [8] had reported five years earlier and has ever since been referred to as the Lorentz transformation (LT), i.e. with a value of $\varepsilon = 1$ (Einstein refers to this function as φ in Ref. 1) in Lorentz's general eqs. (1a-d). Consequently, Larmor's previous conclusion about the FLC also became an integral part of STR, as well as Poincaré's conjecture [7] of remote non-simultaneity.

One of the most interesting features of Einstein's derivation of the LT is that it contains a different justification for choosing a value of unity for ε / φ than Larmor gave. On p. 900 of Ref. 1, he states without further discussion that "φ is a temporarily unknown function of v," the relative speed of the two rest frames S and occurring in the LT. He then goes on to show that considerations of symmetry exclude any other possible value for this function than unity.

Einstein proposed a number of intriguing experiments to test STR, many of which involved the phenomenon of time dilation. The relationship between the time intervals Δt and Δt' between the same two events based on stationary clocks in S and , respectively, can be obtained directly from eq. (1a) as:

$$\Delta t' = \gamma\varepsilon\,(\Delta \mathrm{t} - v\Delta x c^{-2}), \tag{2}$$

whereby Δx is the distance between the locations of the clock in S at the times the measurements are made there. If one has the normal situation in which the two measurements are made at exactly the same location in S, i.e. with $\Delta x = 0$, it follows from STR (with $\varepsilon = 1$) that

$$\mathrm{D}t' = \gamma\Delta t. \tag{3}$$

This equation corresponds to time dilation in the rest frame of the observer in S. It states that the observer in must obtain a larger value for the elapsed time since $\gamma > 1$. One can alter the procedure, however, so that the roles of the two observers are reversed. The inverse relation to eq. (1a) in the LT, i.e. with $\varepsilon = 1$, is:

$$t = \gamma(t' + vx'c^{-2}). \tag{4}$$

The corresponding relation between time intervals thus becomes

$$\Delta t = \gamma\Delta t' \tag{5}$$

when one makes the standard assumption that the measurements are carried out at the same location in ($\Delta x' = 0$). When one compares eq. (3) with eq. (5), it is evident that there is something paradoxical about Einstein's result. This comparison seems to be telling us that each clock can be running slower than the other at the same time. It is always the "moving" clock that is running slower than its identical stationary counterpart, although it is not self-evident how this distinction can be made unequivocally based on the above derivations. Many authors [12-14] have taken this result to be a clear indication that something is fundamentally wrong with Einstein's STR. The majority view among relativistic physicists nonetheless holds that such a situation is a direct consequence of the RP [15], which holds that all inertial systems are equivalent.

The same type of symmetry arises for length measurements, in which case analogous manip- ulations of eq. (1b) and its inverse lead to both:

$$\Delta x = \gamma\Delta x. \tag{6}$$

$$\Delta x = \gamma\Delta x', \tag{7}$$

which are the FLC predictions in the direction parallel to the relative velocity of the two observers. The orthodox interpretation is that length contraction and time dilation occur together in both rest frames, i.e. eq. (5) goes with eq. (6), whereas eq. (3) goes with eq. (7). In other words, lengths on the moving object contract in the parallel direction at the same time (and by the same factor) that the rates of its clocks slow down. There is no difference in the length measurements in directions that are perpendicular to the relative velocity of the two rest frames according to STR [1], as is easily seen from eqs. (1c-d) using the value of ε = 1 which defines Einstein's (and Larmor's) LT. The resulting anisotropic character of the FLC is another of the STR predictions that has evoked much discussion and controversy over the years.

It is important see that the above symmetry in the predicted results of the two observers represents a stark departure from the long-held belief in the objectivity of measurement. Since time-immemorial it had previously been assumed that different observers should always agree in principle as to which piece of cloth is longer or which bag of flour weighs more. The rational system of units is clearly based on this principle. Accordingly, one assumes that while individual observers can disagree on the numerical value of a given physical measurement because they use different units to express their results, they should nonetheless agree completely on the ratio of any two such values of the same type. In what follows, this state of affairs will be referred to as the Principle of Rational Measurement (PRM). Einstein's theory [1] claims instead that observers in relative motion may not only disagree on the ratio of two such measured values but even on their numerical order. It is not feasible to simply state that, on the basis of eq. (6), the reason the observer in S obtains a smaller value for the length of a given object than his counterpart in S' is because he employs a unit of distance which is γ times larger. One could state with equal justification based on eq. (7) that the unit of distance in S is γ times smaller than in . In short, the whole concept of rational units is destroyed by the predicted symmetric character of measurement in Einstein's STR.

Einstein's belief in the above symmetry principle was by no means absolute, however. In the same paper [1], he includes a discussion of a clock that moves from its original rest position and returns via a closed path at some later time. The conclusion is that the rate of the moving clock is decreased during its journey, so that upon return to the point of origin it shows a smaller value for the elapsed time than an identical clock that remained stationary there. The symmetry is broken by the fact that it is possible to distinguish which clock remained at rest in the original inertial system and which did not [1,16]. Einstein ends the discussion by concluding that a clock located at the Equator should run slower by a factor of $\gamma(RE\omega)$ than its identical counterpart at one of the Poles (RE is the Earth's radius and ω is the circular frequency of rotation about the polar axis). This explanation raises a number of questions of its own, however. For example, how can we be sure that a moving clock did not undergo acceleration before attaining its current state of uniform translation? If it did, then its clock rate should be determined independently by its current speed relative to the inertial system from which it was accelerated, and should not depend, at least directly, on its speed relative to a given observer in another inertial system.

The answers to the above questions could only be determined by experiments which were not yet available in 1905. For example, Einstein pointed out that time dilation should produce a transverse (second-order) Doppler effect not expected from the classical theory of light propagation. Eqs. (3) and (5) indicate that a moving light source should emit a lower frequency (greater period) of radiation for a given observer than that of an identical stationary source in his laboratory. The symmetry principle derived from the LT indicates that two observers exchanging light signals should each measure a red shift, i.e a lowering in frequency and corresponding increase in wavelength, from the other's source of radiation when the line joining them is perpendicular to the direction of their relative velocity.

Two years later [17], Einstein used the results of his 1905 paper to derive the Equivalence Principle (EP) between kinematic acceleration and gravity. His analysis of electromagnetic interactions at different gravitational potentials led him to make several more revolutionary predictions on the basis of the EP. First of all, he concluded that the light-speed postulate did not have universal validity. He claimed instead that it should increase with the altitude of the light source. At the same time, he determined that the frequency of light waves should increase as the source moves to a higher gravitational potential. Consequently, light emanating from the Sun should be detected on the Earth's surface with a lower frequency than for an identical source located there. This effect has since been verified many times and in general is referred to as the *gravitational red shift.*

3. Six experiments that led to GPS

3.1. Transverse Doppler effect

The first significant test of Einstein's time-dilation theory was carried out by Ives and Stilwell in 1938 [18]. The object of this experiment is to measure the wavelength λ of light emitted from a moving source at nearly right angles to the observer in the laboratory. In addition to the ordinary first-order Doppler effect, it was expected from STR [1] that a quadratic shift from the normal wavelength (λ_0) should be observed by virtue of time dilation at the source. Because of the constancy of the speed of light and the predicted decrease in frequency emanating from the source, an increase in wavelength should be found in accordance with the transverse Doppler effect according to the general relation:

$$\lambda = \gamma(v)\lambda_0 (1 - vc^{-1} \cos\Theta), \tag{8}$$

(Θ is the angle from which the light is observed). Of the many difficulties inherent in the experiment, the main one was the requirement of measuring the wavelength of light emitted in a perpendicular direction ($\Theta = \frac{\pi}{2}$) relative to the observer. The ingenious solution proposed by the authors was to average the wavelengths obtained for light emitted in both the forward and backward directions, thereby removing the first-order dependence on v in eq. (8). The test was performed with hydrogen canal rays for velocities in the neighborhood of v= 0.005 c. The two shifted Hβ lines in the corresponding spectra were recorded on the same photographic plate as the un-shifted line at 4861Å. The predicted shift in the experiment was 0.0472Å, i.e. $\lambda - \lambda 0$, with v= 0.00441 c. Six different plates were considered and the average shift observed was +0.0468Å. The sign is also in agreement with expectations based on eq. (8), that is, a shift to the red was measured in each case.

The accuracy of the Ives-Stilwell experiment was gradually improved. It was later estimated that the actual experimental uncertainty in the original investigation lay in the 10-15% range, although the experimental points could be fitted to a curve with as little as 2-3% deviation. Mandelberg and Witten [19] made substantial improvements in the overall procedure, using hydrogen ion velocities of up to 0.0093 c. Their results indicate that the exponent in the γ(v) factor in eq. (8) for the quadratic shift has a value of 0.498±0.025, within about 5% of Einstein's predicted value of 0.5.

In spite of the generally good agreement between theory and experiment with regard to the time-dilation effect, there is still a loose end that has been almost universally ignored in the ensuing discussion. It is clear that the predicted slowing down of clocks in the transverse Doppler investigations is accompanied by an increase in wavelength, not the decrease that must be expected on the basis of the FLC (see Sect. II). According to the RP, the wavelength of the radiation that would be measured by an observer co-moving with the light source is just the normal value obtained when the identical source is at rest in the laboratory. The only way to explain this deduction is to assume that the refraction grating that is used in the rest frame of the moving source has *increased in length* by the same factor $\gamma(v)$ as the clock rates have decreased. Moreover, the amount of the increase must be the same in all directions because the clock-rate decrease is clearly independent of the orientation of the radiation to the laboratory observer. That is the only way to explain how the speed of light can be the same in all directions, at least if one believes in an objective theory of measurement. This conclusion is quite different than for the anisotropic length contraction phenomenon predicted by the FLC. *It indicates instead that time dilation in a given rest frame is accompanied by isotropic length expansion.* We will return to this point in the discussion of the next series of time-dilation tests involving the decay of meta-stable particles.

3.2. Lifetimes of meta-stable particles

Shortly after the first Ives-Stilwell investigation, Rossi and coworkers [20] reported on their measurements of the transition rate of cosmic-ray muons and other mesons traveling at different speeds relative to the Earth's atmosphere. The effect of time dilation is much greater in this experiment since the particles are observed to move with speeds exceeding v= 0.9 c. The lifetime τ0 of muons at rest is 2.20×10^{-6} s. The decay is exponential and so the transition probability in the laboratory is proportional to $\exp(-t/\tau_0)$. According to eq. (5) (with $t' = \tau_0$), the lifetime τ of the accelerated muons in the Earth's atmosphere should be $\gamma(v)$ times larger. It was found that the survival probability is proportional to $\exp(-t / \gamma\tau_0)$ for all values of v, which is therefore consistent with the predicted increase in the muon lifetime. Subsequent improvements in the experiment [21] verified this result to within a few per cent.

In the original studies of meson decays in the atmosphere [20], the quantity which was determined experimentally is the average range before decay L. It was assumed that the value of L is proportional to both the speed v of the particles and their lifetime τ in a given rest frame. Consequently, the range is shorter from the vantage point of the particles' rest frame than for an observer on the Earth's surface, since the latter's measured lifetime is longer. Implicit in this conclusion is the assumption that the speed v of the particles relative to the Earth's surface is the same for observers in all rest frames, and thus that L/τ has a constant value for each of them. The results of the latter experiments have subsequently been used in textbooks [22, 23] to illustrate the relationship between time dilation and length contraction in STR [1]. Closer examination of the measured data shows that this conclusion is actually incorrect. The reason that the range of the particles is shorter for the observer moving with the meta-stable particles is clearly because his unit of distance is longer than for his counterpart O on the Earth's surface. The distance traveled is actually the same for both observers, just as is the amount of elapsed time for them. The reason that O measures a longer time is because his clock runs $\gamma(v)$ times faster. Yet both agree on the speed v of the muons based on their respective units of time and distance. This means that the unit of time (his "second") for is $\gamma(v)$ s using the standard definition of 1.0 s employed by O in his rest frame [24]. Similarly, the unit of distance (his "meter") for must also be γ(v) times larger than the standard unit of 1.0 m employed by O. As a result, finds the distance between his position and the Earth's surface to be systematically smaller by a factor of γ(v) than does O. As before with the Ives-Stilwell experiment, the conclusion is thus not that lengths in the rest frame of the accelerated system have contracted, but rather that they have expanded relative to those of identical objects in the Earth's rest frame. Moreover, the amount of the length expansion must be the same in all directions, since otherwise it is impossible to explain how and O can still agree on the value of the speed of light independent of its orientation to each of them.

3.3. High-speed rotor experiments

The above two experiments were in quantitative agreement with Einstein's predictions about the amount of time dilation [1], but neither one of them provided a test for the truly revolu- tionary conclusion that measurement is *subjective*. As discussed in Sect. II, Einstein's LT leads to both eqs. (3) and (5) and therefore is completely ambiguous about which of two clocks in motion is running slower. Yet his argument about clocks located respectively on the Equator and one of the Earth's Poles [1] suggests a completely different, far more traditional (objective), view of such relationships. Both theories indicate that accelerated clocks should run slower than those left behind at their original location. One needs a "two-way experiment" to actually differentiate between the two theoretical positions, one in which the "observer" is moving faster in the laboratory than the object of the measurements

Hay et al. [25] carried out x-ray frequency measurements employing the Mössbauer effect that had the potential of resolving this question. Both the light source and the absorber/detector were mounted on an ultracentrifuge that attained rotational speeds of up to 500 revolutions per second. The authors thereby eliminated the angular dependence in eq. (8) by ensuring that the relative motion of the source to the detector was almost perfectly transverse ($\Theta = \frac{\pi}{2}$). Even more important in the present context, the absorber was fastened near the rim of the rotor while the light source was located close to the rotor's axis [25], which means that the situation is opposite to that in the Ives-Stilwell investigation [18] and it is now the observer who is moving faster in the laboratory.

According to Einstein's LT [1], the above distinction should have no bearing on the outcome of the experiment. This point is borne out by Will's analysis of the transverse Doppler effect [26] in which he expresses the expected result as follows [see his eq. (6)]:

$$\frac{v_r}{v_e} = \left(1 - v_e^2\right)^{0.5}, \tag{9}$$

where v_r and v_e are the observed frequencies at the absorber/receiver and the emitter, respectively, and v_e is the speed of the emitter relative to the receiver. The symmetry in this expression is obvious. The receiver should always measure a smaller frequency than that observed at the light source. Will's analysis is also consistent with the Mansouri-Sexl approach [27] used to detect violations of STR.

In describing their experimental results, Hay et al. [25] state that the "expected shift can be calculated in two ways." By this they mean either by treating the acceleration of the rotor as an effective gravitational field (Einstein's EP) or by "using the time dilatation of special relativity." However, the empirical formula they report for the expected fractional shift in both the energy and frequency of the gamma rays is not consistent with eq. (9) since it is proportional $\frac{(R_1^2-R_2^2)\omega^2}{2c^2}$ where R_1 and R_2 are the respective distances of the absorber and x-ray source from the rotor axis [25]. In other words, the sign of the shift changes when the positions of the absorber and source are exchanged, in clear violation of Einstein's symmetry principle for time dilation. What the empirical formula actually shows is that the magnitude of the effect is correctly predicted by the LT, but not the sign.

The rotor experiments were also carried out later by Kündig [28] and by Champeney et al. [29]. The latter authors report their results in terms of the speeds va and vs of the absorber and the x-ray source, respectively:

$$\frac{\Delta\nu}{\nu}=\frac{v_a^2-v_s^2}{2c^2}. \tag{10}$$

This result makes several additional points clear: a) there is a shift to higher frequency (to the blue) if the absorber is rotating faster than the source and b) the magnitude of the shift is consistent with a higher-order formula:

$$\frac{\Delta\nu}{\nu}=\frac{\gamma(v_a)}{\gamma(v_s)}. \tag{11}$$

If the x-ray source is at rest in the laboratory ($v_s = 0$), eq. (11) reduces to $\frac{\Delta\nu}{\nu}=\gamma(v_a)$. Kündig [28] summarizes this result by stating "that the clock which experiences acceleration is retarded compared to the clock at rest." The slower a clock, the more waves it counts per second [24], hence the observed blue shift when the absorber is located farther from the rotor axis than the source. The empirical results are thus seen to be consistent with Einstein's speculation [1] about the relative rates of clocks located at different latitudes on the Earth's surface. The ambiguity implied by eqs. (3) and (5), as derived from the LT, is replaced by a completely objective theory of time dilation satisfying the relation:

$$\gamma(v_0)\,\Delta t=\gamma(v'_0)\,\Delta t', \tag{12}$$

where v_0 and $_0$ are the respective velocities of the clocks relative to a specific rest frame, which in the present case is the rotor axis. The above equation is also seen to be consistent with both the Ives-Stilwell [18,19] and meta-stable decay experiments [20,21], for which the observer is at rest in the laboratory ($v_0 = 0$) and the object of the measurement has undergone acceleration to speed relative to him.

It is important to see that the alternative derivation [29] of the empirical formula for frequency shifts given in eq. (10) that makes use of Einstein's EP [17] also corresponds to an objective theory of measurement. Accordingly, one assumes the relationship for the gravitational red shift [30]:

$$\frac{\Delta\nu}{\nu}=\frac{-\Delta\varphi}{c^2}=\frac{\omega^2}{2c^2}\left(R_a^2-R_s^2\right)=\frac{v_a^2-v_s^2}{2c^2} \tag{13}$$

where the centrifugal field of force corresponds to a radial gravitational field strength of $R\omega^2$ in each case ($\Delta\varphi$ is the difference in gravitational potential). There is clearly no ambiguity as to which clock lies higher in the gravitational field and thus there is no question as to which clock runs slower in this formulation. This point is missed in many theoretical descriptions of the frequency shifts in the rotor experiments. For example, Sard [30] claims that the gravita- tional red shift and the transverse Doppler effects are consistent, and that this "is not surpris- ing, since both follow in a straightforward way from the interweaving of space and time in the Lorentz transformation." This view completely overlooks the simple fact that the transverse Doppler formula in eq. (9) derived from the LT does not allow for the occurrence of blue shifts, whereas the corresponding EP result in eq. (13) demands that they occur whenever the absorber in the rotor experiments moves faster than the x-ray source relative to the laboratory rest frame. Sherwin [31] attempts to clarify the situation by asserting that the ambiguity inherent in eq. (9) only holds for uniform translation, but he fails to give actual experimental results that verify this conclusion. Both Rindler [32] and Sard [30] try to justify the empirical result in terms of orthodox STR by considering the relative retardations experienced by the two clocks in the rotor experiments. This argument leads directly to eqs. (10-11), *but it does so by eliminating the basic subjective character of the LT in doing so*. Once one assumes that two clock rates are strictly proportional, in accord with eq. (12), there is no longer any basis for claiming that the measurement process is subjective. More discussion of this general subject may be found in Ref. [33].

3.4. Terrestrial measurements of the gravitational red shift

The gravitational red shift derived in Einstein's Jahrbuch review [17] is expressed in the equation below [cf. eq. (13)]:

$$\nu_D = \nu_x(1 + \Phi c^{-2}). \tag{14}$$

In this formula, ν_X is the frequency of light emitted from a source that is at rest at a given gravitational potential, whereas νD is the corresponding value of the frequency detected by an observer who is not in relative motion to the source but is located at some other position in the gravitational field. The potential energy difference Φ between the source and detector determines the fractional amount of the frequency shift. For an experiment near the Earth's surface, $\Phi = gh$, where h is the difference in altitude between the source and detector and g is the acceleration due to gravity. The sign convention is such that $\Phi > 0$ when the source is located at a higher gravitational potential than the detector.

The same proportionality factor occurs [17,34] in the relationship between the light speeds measured at the same two positions in the gravitational field, where c_D is the value measured at the detector and c_X is the corresponding value measured at the light source:

$$c_D = c_X\left(1 + \Phi c^{-2}\right). \tag{15}$$

As a result, it is clear that according to the EP, both the frequency and the speed of light increase as the light source moves to a higher gravitational potential. Moreover, the fractional amount of the change is the same in both cases. As will be discussed in the following, there is ample experimental evidence to indicate that both of these relations are correct.

Terrestrial confirmation of the gravitational red shift first became possible with the advent of the Mössbauer technique for detecting changes in x-ray frequencies. Pound and Snider [35] placed a ^{57}Fe source at a distance h of 22.5 m above an absorber. According to eq. (14), the fractional change in the x-ray frequency ($3.47x10^{18}$Hz) should have a value of $ghc^{-2}=2.45x10^{-15}$. The authors employed the EP directly to obtain their results. By imparting a downward velocity of $v= - ghc^{-1}$ to the detector, the resulting first-order Doppler effect exactly balanced the gravitational red shift:

$$\nu_D = \nu_X\left(1 + vc^{-1}\right). \tag{16}$$

The observed results were found to be within expectations by a factor of 0.9990±0.0076 [35].

The interpretation of the above experiment is nonetheless a matter of some confusion. Pound and Snider [35] basically avoided a theoretical discussion of the ramifications of their work vis-ŕ-vis the general theory of relativity (GTR [36]), but they noted that "the description of the effect as an 'apparent weight' of photons is suggestive. The velocity difference predicted is identical to that which a material object would acquire in free fall for a time equal to the time of flight." The latter value is assumed to be $t= hc^{-1}$, so that the expected velocity increase of the falling particle would indeed be $\Delta v= gt = ghc^{-1}$. However, it is important to see that the frequency of the emitted light in *absolute terms* is actually the same at the light source as it is on the ground below when it arrives there. As Einstein stated in his 1911 article [34], this is a clear physical requirement because the number of wave crests arriving at a given point over a specific time interval is a constant [37]. *The reason that the detector at the ground level records a higher frequency is because the unit of time there is greater than at the location of the light source.* Therefore, more wave crests are counted in 1 s on the ground clock than for its counterpart at a higher altitude. Comparison of the rates of atomic clocks located at different altitudes over long periods of time has confirmed Einstein's conclusion quantitatively on this point [38].

The above discussion raises another question, however, namely what happens to the speed of light as it descends from the source? According to eq. (15) and Einstein's EP [17,34], the answer is clearly that it decreases, and by the same fractional amount as for the associated frequency red shift given in eq. (14). To be specific for the Pound-Snider example, the speed of light must have a value of c at ground level for an observer located there, whereas the same observer measures the value at the light source to be ghc^{-2} times larger. In other words, the speed of light actually decreases as it passes downward from the source to the detector at ground level. This is clearly the opposite change that is expected from the "apparent weight" argument alluded to above. Again, there is ample experimental evidence that Einstein's eq. (15) is quantitatively accurate. One knows, for example, from the work of Shapiro et al. [39] that radio signals slow in the neighborhood of massive bodies such as Venus and Mercury. There are time delays between the emission of radio pulses towards either of these planets and detection of their echoes when they return to the Earth's surface. This shows that the speed of light decreases under the influence of a gravitational field, i.e. with $\Phi < 0$ in eq. (15).

There is another important question that needs to be considered in the present discussion, however, namely what happens to the *wavelength of light* as the source changes its position in a gravitational field? Einstein [17] concluded that "light coming from the solar surface… has a longer wavelength than the light generated terrestrially from the same material on Earth [40]." Since this statement comes directly after his proof of eq. (14) for frequencies, it seems quite plausible he was basing it on the assumption that the product of wavelength λ and frequency ν is equal to the speed of light. In order to obtain the quoted result on this basis, however, it was also necessary for him to assume that the speed of light is actually the same at the Sun's surface as on the Earth. This explanation is nonetheless in conflict with eq. (15), which it must be emphasized appears in the same publication [17]. Making the same assumption about the relationship between light speed, wavelength and frequency and comparing eq. (15) with eq. (14) leads unequivocally to the conclusion that the *wavelength of light for a given source is completely independent of the latter's position in a gravitational field*:

$$\lambda_D = c_D \nu_D^{-1} = c_X \nu_X^{-1} = \lambda_X. \tag{17}$$

It is important in general to recognize that in each of eqs. (14,15,17), the same quantity is the object of measurement for two different observers. The only reason why these observers do not obtain the same result in each case is *because they employ different units for their various measurements.* In Sect. II, this state of affairs has been referred to as the Principle of Rational Measurement (PRM). The quantity in parentheses in eqs. (14,15), $(1 + \Phi c^{-2})$, can be looked upon as a conversion factor between their respective units; it will be designated in the ensuing discussion simply as S. It will be seen that the conversion factors for other physical quantities can always be expressed as powers of S. For example, the conversion factor for wavelength, and therefore distances in general because *the scaling must be completely uniform in each rest frame*, is S^0. These conversion factors satisfy the ordinary rules of algebra. For example, since frequency and time are inversely related, it follows that the conversion factor for times is S^{-1}.

The unit of velocity or speed is equal to the ratio of the distance travelled to elapsed time and hence the conversion factor in this case is equal to the ratio of the corresponding conversion factors, namely $S^0/S^{-1} = S$, consistent with the previous definition from eq. (15).

It is useful to employ the above concepts to follow the course of the light pulses in the Pound- Snider experiment [35]. The analysis of two events is sufficient to illustrate the main points in this discussion, namely the initial emission of the x-rays at the higher gravitational potential (I) and their subsequent arrival at the absorber (II). It is important to use the same set of units in comparing the corresponding measured results. In terms of the units at the x-ray source, the observer X there measures the following values for event I: $\nu_X(I) = \nu$, $\lambda_X(I) = \lambda$ and $c_X(I) = c$, where ν and λ are the standard frequency and wavelength of the light source. Upon arrival at the absorber, the same observer measures the corresponding values: $\nu_X(II) = \nu$, $\lambda_X(II) = \lambda S^{-1}$ and $c_X(II) = cS^{-1}$. In other words, the value of the frequency has not changed during the passage of the x-rays down to the absorber, *but the speed of light has decreased* in accordance with the EP (S>1). Consequently, the wavelength of the light must have decreased by the same factor in order to satisfy the general relation (phase velocity of light) in free space between frequency, wavelength and light speed ($\lambda_X \nu_X = c_X$).

The observer D located at the absorber measures generally different values for the same two events, not because he is considering fundamentally different processes, but rather because he bases his numerical results on a different system of physical units (PRM). He therefore finds for event II: $\nu_D(II) = S\nu$, $\lambda_D(II) = \lambda S^{-1}$ and $c_D(II) = c$. His unit of distance is the same and so there is complete agreement on the value of the wavelength (λS^{-1}). However, his units of frequency and speed are S times smaller than for his counterpart at the higher gravitational potential, and therefore he measures values for these two quantities which are S times *greater* in each case. This set of results illustrates the very important general principle of measurement, namely *that the numerical value obtained is inversely proportional to the unit in which it is expressed.* The corresponding values for the initial emission process are accordingly: $\nu_D(I) = S\nu$, $\lambda_D(I) = \lambda$ and $c_D(I) = Sc$.

In absolute terms what has happened as a result of the downward passage of the x-rays between source and absorber/detector is that the light frequency has remained constant throughout the entire process, exactly as Einstein demanded in his Jahrbuch review [17]. On the other hand, both the light speed and the corresponding wavelength of the radiation have deceased by a factor of S. This is in agreement with Einstein's second postulate of STR [1], namely that the observer at the absorber must find that the speed of light when it arrives there has a value of c in his units because it is at the same gravitational potential as the observer at that point in time.

Before closing this section, it is worthwhile to mention how the units of other physical properties vary/scale with gravitational potential. To begin with, energy E satisfies the same proportionality relationship as for frequency and light speed [see eqs. (14,15)]:

$$E_D = E_X\left(1 + \Phi c^{-2}\right) = SE_X. \tag{18}$$

Indeed, one can derive this equation to a suitable approximation using Newton's gravitational theory and the definition of gravitational potential energy as mgh=mΦ. According to STR [1], the value of the energy EX at the light source is equal to mc2 and thus the

fractional increase is $ghc^{-2} = \Phi c^{-2}$, in agreement with eq. (18) at least for infinitesimal values of h for which the difference between inertial and gravitational mass is negligible. *The fact that energy and frequency scale in the same way means that Planck's constant $\hbar$ does not vary with gravitational potential.* This in turn means that angular momentum in general scales as S0, the same as for distance.

Einstein also stated in his Jahrbuch article [17] that the above dependence of energy on gravitational potential implies that there is a position-dependent component corresponding to an inertial mass m equal to E/c^2 [40]. Because E and c [i.e. the generic speed of light of eq. (15)] both vary as S, it follows that the unit of inertial mass scales as S-1, the same as time. Values of linear momentum p=mv are the same for observers located at different gravitational potentials, since the corresponding scaling/conversion factors for inertial mass and velocity exactly cancel one another. With reference to the Pound-Snider experiment [35], this means that x-ray photons *increase in momentum* as they fall through the gravitational field. This follows from quantum mechanics and the fact that the corresponding wavelength of the radiation decreases in the process. Hence, the E= pc relation holds for observers located at all gravitational potentials since the light speed is decreasing at exactly the same rate as the momentum is increasing, while the energy of the photons is conserved throughout the entire process. This situation might seem counter-intuitive, but it emphasizes that photons are different than material particles with non-zero rest mass, for which momentum and velocity are strictly proportional (see also the related discussion of light refraction in Ref. 41). Note also that because of the governing quantum mechanical equations, this result is consistent with the $\nu\lambda$ =c relation for light waves, also already alluded to above. More details about gravitational scaling may be found in a companion publication [42].

3.5. Clock rates on circumnavigating airplanes

Despite the progress that had been made in carrying out experiments that verify various aspects of Einstein's STR [1], there was still considerable uncertainty as to how to predict the results of future investigations of the time-dilation effect by the time the next significant advance was made in 1971. At that time, Hafele and Keating [43-44] carried out tests of cesium atomic clocks located onboard circumnavigating airplanes that traveled in opposite directions around the globe. In their introductory remarks, the authors noted that the "relativistic clock paradox" was still a subject of "the most enduring scientific debates of this century." The first paper presents some predictions of the expected time differences for clocks located on each airplane as well as at the origin of their respective flights. Their calculations are clearly based on an objective theory of time dilation that is consistent with eq. (12). The symmetric relationship that is inherent in eqs. (3) and (5) derived from the LT is completely ignored. Gone is the idea that the observer at the airport should find that the clocks on the airplanes run slower while his counterpart on the airplane should find the opposite ordering in clock rates. To sustain this argument, Hafele and Keating argue [43] that standard clocks located on the Earth's surface are generally not suitable as reference in their calculations because of their acceleration around the Earth's polar axis.

All speeds to be inserted into eq. (12) must therefore be computed relative to the Earth's non- rotating polar axis, or more simply, relative to the gravitational midpoint of the Earth itself. The latter therefore serves the same purpose as the rotor axis in the Hay et al. experiments [25] discussed in Sect. III.C. In short, the rotational speed $R\Omega \cos\chi$ of the Earth at a given latitude χ must be taken into account for both objects on the ground and on the airplanes. If the airplane travels with ground speed v in an easterly direction, this means that the ratio of the elapsed times of an onboard clock τ and that of a clock on the ground τ_0 is:

$$\frac{\tau}{\tau_0} = \frac{\gamma(R\Omega\cos\chi)}{\gamma(v + R\Omega\cos\chi)} \sim \left|1 - \frac{2vR\Omega\cos\chi + v^2}{2c^2}\right. . \tag{19}$$

The interesting conclusion is that a clock traveling in the westerly direction with the same ground speed (v< 0) should run faster than both its easterly counterpart and also the ground clock at the airport of departure.

The latter expression does not include the effect of the gravitational red shift discussed in Sect. III.D. If the altitude of the airplane relative to the ground is h, the ratio becomes:

$$\frac{\tau}{\tau_0} \sim 1 + \frac{gh}{c^2} - \frac{2vR\Omega \cos\chi + v^2}{2c^2}. \tag{20}$$

The gravitational increase in clock rate on the airplane was typically somewhat smaller than the time-dilation effect for the clock traveling eastward, so that the overall effect was still a decrease in its rate relative to the clock at the airport. The two effects reinforce each other for the airplane traveling in the westerly direction.

The predicted effects were mirrored in the experimental data [44]. The observed time difference between the ground and easterly traveling clocks was -40 ± 23 ns as compared to the predicted value of -59 ± 10 ns. The corresponding values for the westward traveling clock were +275 ± 21 ns vs. 273 ± 7 ns. The main sources of error were due to instabilities in the cesium clock rates and uncertainties in the actual flight path of the airplanes.

3.6. Transverse Doppler effect on a rocket

The transverse Doppler effect was again used to study the effects of motion and gravity on clock rates in an experiment carried out in 1976 by Vessot and Levine [45]. The frequency of a hydrogen-maser oscillator system carried onboard a rocket was measured as it moved up and down in the Earth's gravitational field. The first-order Doppler effect was eliminated by using a transponder system on the rocket, leaving only the second-order transverse Doppler effect and the gravitational red shifts as the cause of the observed frequency shifts. The flight lasted for nearly two hours and carried the rocket to a maximum altitude of 10000 km.

The speed of the rocket was monitored as a function of altitude and the amount of the frequency shift was predicted on the basis of eq. (20) except that one took into account the variation of g with altitude to obtain the desired accuracy and also consider the effect of the centrifugal acceleration of the ground station [see eq. (1) of Ref. 45]. The position and velocity of both the rocket and the ground station were measured in an Earth-centered inertial frame, consistent with the assumptions of the Hafele-Keating analysis [43] of their data for circumnavigating airplanes. It was also necessary to take account of the refractive index of the ionosphere as the rocket passed through this region of space. The Doppler-canceling feature of the experiment was also successful in removing the gross effects of such variations.

The authors concluded that the rocket experiment was consistent with Einstein's EP to a suitably high level of accuracy. By assuming the validity of the EP, they were also able to place an upper bound on the ratio of the one-way and two-way velocity of light. The corresponding $\Delta c/c$ values ranged from $1.9\text{x}10^{-8}$ to $-5.6\text{x}10^{-8}$ during various parts of the rocket's trajectory.

4. Application of the experimental clock-rate results

The experiments described in the preceding question provide the necessary empirical data to enable the accurate prediction of how the rates of clocks vary as they are carried onboard rockets and satellites. This information is critical for the success of the Global Positioning System because of the need to measure the elapsed times of light signals to pass between its satellites and the Earth's surface.

Before considering how to make the desired predictions based on the above experience, it is interesting to examine how Einstein's original theory of clock rate variations [1] has held up in the light of actual experiments. There has been unanimity in the theoretical discussions accompanying these experiments in claiming that their results mirror perfectly Einstein's predictions. This is particularly true of the authors of the transverse Doppler experiments using high-speed rotors [25,28-29] as well as for Hafele and Keating in their investigations of the rates of atomic clocks onboard circumnavigating airplanes [43,44].

Such an evaluation overlooks the basic point discussed in Sect. II, however, namely that Einstein had two clearly distinguishable conclusions regarding the question of which of two identical clocks in relative motion runs slower in a given experiment [1]. One set of predictions was based on the LT and the conclusion that all inertial systems are equivalent by virtue of the RP. This is a thoroughly subjective theory of measurement which is characterized by the ambiguous, seemingly contradictory, results obtained in eqs. (3) and (5), both of which are derived in a straightforward manner from the LT. In this view, which clock runs slower is simply a matter of one's perspective. This conclusion is consistent with Will's equation [26] for the transverse Doppler effect (see eq. [9]), which states that the emitted frequency of a moving light source is necessarily lower than for the same source when it is at rest in the laboratory of the observer. By contrast, Einstein's alternative conclusion was that an acceler- ated clock is definitely slowed relative to one that remains at rest in its original location. That corresponds to a thoroughly objective theory of measurement consistent with eq. (12), which requires the assignment of a definite origin (objective rest system ORS [46]) from which to measure the speeds of the respective clocks being compared.

There is no doubt which of Einstein's two conclusions is meant by the various authors when they assert that their results are perfectly consistent with relativity theory. This point was made quite clear in Sherwin's analysis [31] of the rotor experiments, which he referred to as a demonstration of "the transverse Doppler effect for accelerating systems." It is the unambiguous nature of the result of the clock paradox that sets the rotor experiments apart from the classical theory of time dilation derived from the LT. Hafele and Keating [43] take the same position in presenting the underlying theory of their measurements of the rates of atomic clocks. In that case, the authors claim the determining factor for the choice of the Earth's non- rotating axis as the reference system for application of eq. (12) is that it is an inertial frame. However, this conclusion overlooks the fact that the Earth is constantly accelerating as it makes its way around the Sun, and thus it is at least quantitatively inaccurate to claim that a position at one of the Earth's Poles is in constant uniform translation [46].

A simpler alternative view is that the Earth's center of mass (ECM) plays the same role for clock rates as for falling bodies. The only way to change the direction in which free fall occurs is to somehow escape the gravitational pull of the Earth. This has been most notably accomplished with the NASA expeditions to the Moon, in which case there is a definite point in the journey where the rocket tends to fall away from planet Earth and towards its satellite instead. It also occurs in a more conventional fashion when a body undergoes sufficient centrifugal acceleration in a parallel direction to the Earth's surface.

In this view, the ORS [46] for computing time dilation is always the same as the reference point for free fall, and thus is subject to change in exactly the same manner. The current speed of the Earth relative to the Sun is therefore irrelevant for the quantitative prediction of clock rates for the applications at hand.

There is a far less subtle point to be considered in the present discussion, however. The conclusion that an accelerated clock always runs slower than one that remains at its original rest location has very definite consequences as far as relativity theory is concerned. It shows unequivocally that Einstein's symmetry principle [1] is violated, and therefore that the LT is contradicted by the experimental findings. Discussions in the literature [30-32] always emphasize that the acceleration effect was predicted by Einstein in his original paper [1] and that it is also entirely consistent with his EP [17], but there is never any mention of the conflict with the LT, which is after all the essential cornerstone of STR from which many long-accepted conclusions follow. The latter include most especially the FLC and remote non-simultaneity. Showing that the LT fails in one of its most basic predictions, the ambiguity in the ordering of local clock rates for two moving observers [31], necessarily raises equally critical questions about the validity of all its other conclusions. For example, does Lorentz invariance hold for the space and time variables under these experimental conditions? Are two events that occur at exactly the same time for one observer not simultaneous for someone else who happens to be moving relative to him? The latter question is particularly important for the GPS methodology since it relies on the assumption that the time of emission of a light signal is the same on the ground as on an orbiting satellite as long as one corrects for differences in clock rates at the respective locations. We will return to these questions in Sect. V, but first attention will be focused on the way in which the available experimental data on clock rate variations can be applied in practice.

4.1. Synchronization of world-wide clock networks

There are two causes of clock-rate variations in terrestrial experiments. Both of them involve the ECM, one depending exclusively on the speed of a given clock relative to that position and the other on the corresponding difference in gravitational potential. As a result, in comparing different clocks on the Earth's surface, it is necessary to know both the latitude χ of each clock as well as its altitude h relative to sea level. The time-dilation effect is governed exclusively by eq. (12), whereby the speed to be inserted in the respective γ factors is equal to $R_E \Omega cos\chi$, where Ω is the Earth's rotational frequency (2π radians per 24 h = 86 400 s) and R_E is the Earth's radius (or more accurately, the distance between the location of the clock and the ECM). A standard clock S needs to be designated. Theoretically, there is no restriction on its location. Its latitude χ_S (as well as its altitude rS relative to the ECM) is then an important parameter in computing the ratio of the rates of each clock in the network with that of the standard clock. It is helpful to define the ratio R as follows:

$$R = \frac{\gamma(R_E \Omega \cos\chi)}{\gamma(R_E \Omega \cos\chi_S)}. \tag{21}$$

According to eq. (12), this ratio tells us how much slower (if R> 1) or faster (if R< 1) the given clock runs than the standard if both are located at the same gravitational potential. The gravitational red shift needs to be taken into account to obtain the actual clock-rate ratio. For this purpose, it is helpful to define a second ratio S for each clock:

$$S = 1 + \frac{g(r - r_S)}{c^2}, \tag{22}$$

where r is the distance of the clock to the ECM. This ratio tells us how much faster (S> 1) the clock runs relative to the standard by virtue of their difference in gravitational potential. The elapsed time Δt on the clock for a given event can then be converted to the corresponding elapsed time Δt_S on the standard clock by combining the two ratios as follows:

$$\Delta t_S = \left(\frac{R}{S}\right)\Delta t. \tag{23}$$

It is possible to obtain the above ratios without having any communication between the laboratories that house the respective clocks. The necessary synchronization can begin by sending a light signal directly from the position of the clock A lying closest to the standard clock S. The corresponding distance can be determined to as high an accuracy as possible using GPS. Division by c then gives the elapsed time for the one-way travel of the signal based on the standard clock S. The time of arrival on the standard clock is then adjusted backward by this amount to give the time of emission $T^{S0}(A)$ for the signal, again based on the standard clock. The corresponding time of the initial emission read from the local clock is also stored with the value T0(A). In principle, all subsequent timings can be determined by subtracting T0(A) from the current reading on clock A to obtain Δt to be inserted in eq. (23). The time TS of the event on the standard clock is then computed to be:

$$T^S = \frac{R}{S}\Delta t + T^{S0}(A), \tag{24}$$

where R and S are the specific values of the ratios computed above for clock A.

Once the above procedure has been applied to clock A, it attains equivalent status as a standard. The next step therefore can be applied to the clock which is nearest either to clock A or clock S. In this way the network of standard clocks can be extended indefinitely across the globe. Making use of the "secondary" standard (A) naturally implies that all timings there are based on its adjusted readings. It is important to understand that no physical adjustments need to be made to the secondary clock, rather its direct readings are simply combined with the R and S factors in eq. (24) to obtain the timing results for a hypothetical standard. A discussion of this general point has been given earlier by Van Flandern [47]. The situation is entirely analogous to having a clock in one's household that runs systematically slower than the standard rate. One can nonetheless obtain accurate timings by multiplying the readings from the faulty clock by an appropriate factor and keeping track of the time that has elapsed since it was last set to the correct time. The key word in this discussion is "systematic." If the error is always of quantitatively reliable magnitude, the faulty clock can replace the standard without making any repairs.

4.2. Adjustment of satellite clocks

The same principles used to standardize clock rates on the Earth's surface can also be applied for adjusting satellite clocks. Assume that the clock is running at the standard rate prior to launch and is perfectly synchronized with the standard clock (i.e. as adjusted at the local position). In order to illustrate the principles involved, the gravitational effects of other objects in the neighborhood of the satellite are neglected in the following discussion, as well as inhomogeneous characteristics of the Earth's gravitational field.

The main difference relative to the previous example is that the R and S factors needed to make the adjustment from local to standard clock rate using eqs. (23-24) are no longer constant. Their computation requires a precise knowledge of the trajectory of the satellite, specifically the current value of its speed v and altitude r relative to the ECM. The acceleration due to gravity changes in flight and so the ratio S also has to be computed in a more fundamental way. For this purpose, it is helpful to define the following quantity connected with the gravitational potential, which is a generalization of the quantity in parentheses in eq. (18):

$$A(r) = 1 + \frac{GM_E}{c^2 r}, \tag{25}$$

where G is the universal gravitational constant (6.670×10^{-11} Nm^2/kg^2) and M_E is the gravitational mass of the Earth (5.975×10^{24} kg). The value of S is therefore given as the ratio of the A values for the satellite and the standard clock:

$$S = \frac{A(r_S)}{A(r)}, \tag{26}$$

which simplifies to eq. (22) near the Earth's surface (with $g = \frac{GM_E}{r_S^2}$).

The corresponding value of the R ratio is at least simple in form:

$$R = \frac{\gamma(v)}{\gamma(v_S)} = \frac{\gamma(v)}{\gamma(R_E \Omega \cos \chi_S)}. \tag{27}$$

Note that the latitude χ_S is not that of the launch position relative to the ECM, but rather that of the original standard clock. The accuracy of the adjustment procedure depends primarily on the determination of the satellite speed v relative to the ECM at each instant.

In this application the underlying principle is to adjust the satellite clock rate to the corresponding standard value over the entire flight, including the period after orbit has been achieved. The correction is made continuously in small intervals by using eq. (23) and the current values of R and S in each step. The result is tantamount to having the standard clock running at its normal rate on the satellite. This above procedure super-cedes the "pre- correction" technique commonly discussed in the literature [3] according to which the satellite clock is physically adjusted prior to launch. The latter's goal is to approximately correct for the estimated change in clock rate expected if the satellite ultimately travels in a constant circular trajectory once it achieves orbital speed. The present theoretical procedure has the advantage of being able to account for departures from a perfectly circular orbit and also for rate changes occurring during the launch phase.

5. The new Lorentz transformation

Although the relativistic principles for adjusting the rates of moving clocks are well under- stood, there is still an open question about how this all fits in with Einstein's STR [1]. The standard argument, starting with the high-speed rotor experiments of Hay et al. [25], has been that one only needs to know the speed of the clock relative to some specific inertial system in order to compute the decrease in its rate using eq. (12). While this is true, the procedure itself cannot be said to be consistent with the basic subjectivity of Einstein's original theory. As discussed in the beginning of Sect. IV, one has to disregard the predictions of the LT in order to have a theory which unambiguously states which of two clocks is running slower, namely the one that is accelerated by the greater amount relative to the aforementioned inertial system [28,31]. The consequences of ignoring the predictions of the LT in this situation are generally avoided in theoretical discussions of clock rates, with authors [16, 30-32] preferring instead to explain how correct predictions can be obtained by assuming that the amount of time dilation is proportional to the $\gamma(v)$ factor computed from the speed v of the accelerated clock. The argument is made that the LT can only be used for clocks in uniform translation [31], but the implication is that its many other predictions still retain their validity, such as remote non- simultaneity and Lorentz space-time invariance. However, the fact is that there is no basis for making the latter conclusion. The best that can be said is that there is no essential connection between the two types of predictions, and so failing in one of them does not necessarily rule out the validity of the others.

The standard procedure for addressing such problems is to find a way to amend the theory that removes the contradiction at hand, namely in the present case, the prediction of STR that two clocks can both be running slower than the other at the same time [1], while still retaining the capability of making all the other hitherto successful predictions of the theory without further modification. In previous work [48-50] it has been shown that the following space-time transformation accomplishes this objective by making a different choice for the function ε in eqs. (1a-d) than the value of unity assumed by Einstein in his original derivation of the LT [1]:

$$
\begin{aligned}
t' &= Q^{-1}t && \text{a}\\
x' &= \eta Q^{-1}\left(x - vt\right) && \text{b}\\
y' &= \eta(\gamma Q)^{-1}y && \text{c}\\
z' &= \eta\left(\gamma Q\right)^{-1}z\,. && \text{d}
\end{aligned}
\qquad (28)
$$

In this set of equations, $\eta = (1 - vu_x c^{-2})^{-1} = (1 - vxt^{-1}c^{-2})^{-1}$ and Q is a proportionality factor that defines the ratio of the two clock rates in question. It is assumed that the rates of moving clocks are strictly proportional to one another, thereby eliminating the space-time mixing inherent in the LT. Reviewing the arguments that Einstein used in his derivation of the LT shows that he in fact made an undeclared assumption therein without any attempt to justify it on the basis of experimental observations. The assumption was that ε can only depend on v, the relative speed of the primed and unprimed coordinate systems related in the transformation. It leads to the $y = y'$ and $z = z'$ relations of the LT instead of eqs. (28c-d), but it also forces the mixing of space and time variables in the LT instead of the simpler result in eq. (28a).

The alternative Lorentz transformation (ALT) in eqs. (28a-d) satisfies both of Einstein's postulates [1] because of its direct relation to the general equations in eqs. (1a-d), and it is also consistent with the same velocity transformation (VT) that is an integral part of STR:

$$u'_x = \left(1 - \mathrm{v}u_x c^{-2}\right)^{-1}\left(u_x - \mathrm{v}\right) = \eta\left(u_x - \mathrm{v}\right) \quad \text{a}$$

$$u'_y = \gamma^{-1}\left(1 - \mathrm{v}u_x c^{-2}\right)^{-1} u_y = \eta\gamma^{-1}u_y \quad \text{b}$$

$$u'_z = \gamma^{-1}\left(1 - \mathrm{v}u_x c^{-2}\right)^{-1} u_z = \eta\gamma^{-1}u_z. \quad \text{c}$$

(29)

It is more than an alternative to Einstein's LT, however, because it is possible to show that the LT is invalid even for uniformly translating systems [51]. This can be seen by considering the example of two observers measuring the length of a line that is oriented perpendicularly to their relative velocity. The FLC, which is derived from the LT, predicts [1] that they must agree on this value ($\Delta y = \Delta y'$). If their clocks run at different rates, however, they will measure different elapsed times ($\Delta t \neq \Delta t'$) for a light pulse to travel the same distance. Since they also must agree on the speed of light, however, one concludes that the corresponding two distance measurements also differ, since $\Delta y = c\Delta t \neq c\Delta t' = \Delta y'$.

This contradiction has been referred to as the "clock riddle" in earlier work [50-51], as opposed to the "clock paradox" frequently discussed in the literature. In fact, the $\Delta y = \Delta y'$ prediction is merely the result of Einstein's original (undeclared) assumption mentioned above. The contradiction is easily removed by making a different assumption for the value of ε in eqs. (1a-d). The ALT makes this choice by assuming that the clock rates of the two observers must be strictly proportional, as expressed by eq. (28a). The latter assumption is clearly consistent with the high-speed rotor [25,28-29] and Hafele-Keating [43-44] experiments, and leads to the result expected on the basis of time dilation and the light-speed constancy postulate. The ALT also leads directly to the key experimental results connected with the aberration of starlight at the zenith and the Fresnel light-drag experiment, since they can be derived quite simply from the VT of eqs. (29a-c).

The ALT is the centerpiece of a thoroughly objective version of relativity theory. It subscribes to the Principle of Rational Measurement (PRM), whereby two observers must always agree in principle on the ratio of physical quantities of the same type [42]. There is never any question about which clock is running slower or which distance is shorter, unlike the case for Einstein's STR. The new theory [48,49] does this by restating the RP: The laws of physics are the same in all inertial systems, *but the units on which they are expressed may vary systematically between rest frames.* The passengers locked below deck in Galileo's ship cannot notice differences between traveling on a calm sea and being tied up on the dock next to the shore, but this does not preclude the possibility that all their clocks run at a different rate in one state of constant motion than in the other. Such distinctions first become clear when they are able to look out of a window and compare their measurements of the same quantity with those of their counterpart in another rest frame.

At the same time, the ALT insists on remote simultaneity by virtue of eq. (28a), since it is no longer possible for Δt to vanish without $\Delta t'$ doing so as well. It does away with space-time mixing, the FLC and Lorentz invariance, and it also removes any possibility of time reversal ($\frac{\Delta t}{\Delta t'} < 0$) since the proportionality factor Q is necessarily greater than zero in eqs. (28a-d). Not one of the latter effects has ever been verified experimentally, although there has been endless speculation about each of them. The ALT might be more aptly referred to as the "GPS LT," because it conforms to the basic assumption of this technology that the rates of atomic clocks in relative motion are only affected by their current state of motion and position in a gravitational field. More details concerning these aspects of relativity theory may be found in a companion article [42].

6. GPS positioning errors during strong solar activity

The relativistic and gravitational effects discussed above are observed at sufficiently large distances from GPS satellites to the E-layer of the atmosphere. In accordance with existing experimental data these effects lead to positioning errors which do not exceed 3 m on the Earth's surface. The next largest sources of error may be due to satellite geometry with respect to the GPS receiver, as well as orbit deformation due to the influence of gravity and the uneven distribution of this field.

We consider below another key aspect of the process of point-location determination by GPS. It is based on the fact that the distance can be determined using information from the atomic clock, and the assumption that the light pulses are always moving with the same speed through space. Then, to determine the appropriate distance it is sufficient to multiply this speed by the time of passage of the light pulse between two points in space. Use of this simple approach leads to problems in GPS positioning that arise from the possibility of light pulse propagation at different speeds through free space. This becomes particularly evident during periods of solar activity and the occurrence of magnetic storms, when the positioning errors are quite large. If one takes the generally accepted view that associates the distortion of GPS signals with wave optics, the positioning errors that occur during these periods correspond to an increase in the optical length of the signal propagation. Allowance for refraction in the scattering of waves on the plasma-type seals like blobs and bubbles with refractive index nf = 1.0000162 [52] in the F- and upper E-layers of the ionosphere (at altitudes of 100-400 km above the Earth's surface) leads to errors in positioning which are close to relativistic ones. The latter is in good agreement with experiment [53].

With increasing solar activity the time of the GPS signal passage from satellite to the Earth rises, which leads to positioning error enlargement. And it can be realized both in a short period of time (with duration 5-20 min) and for a long period (lasting several hours). In the first case, it occurs under the influence of the radiation coming from the solar flare. In the second case, it takes place in 30-35 hours after the flare under the influence of the solar wind. A concrete example is the time-dependence of the violations of the GPS satellite system during periods of solar activity which was published on the website of Cornell University [54]. According to measurements carried out in real-time monitoring stations of Arecibo Observatory (Puerto Rico) daily from August 30 to September 02, 2011 between 03.00 and 04.00 on the Coordinated Universal Time (UTC), there was a 20-minute failure of GPS. The horizontal positioning error here reached 50 m and more.

More powerful geomagnetic disturbances lead to the complete disappearance of the signal at the GPS receiver for a sufficiently long period of time [55]. Thus, the data obtained at the Sao Luis Observatory (Brazil) on September 15-16, 2011 showed that the loss of GPS signal occurred several times during the day. The signal at the receiver sporadically disappeared five times for 5-30 minutes each between 16.00 UTC September 15 to 01.00 UTC September 16, 2011. Moreover, the horizontal positioning error during these days greatly exceeded the value of 50 m.

It was shown in [56] that during solar flares of different power there was a certain sequence of decreasing carrier/noise ratio for the frequencies L1=1.57542 GHz and L2=1.22760 GHz. During the solar flare of X-1 level (22.15 UTC, December 14, 2006) the carrier/noise ratio for the frequency L1 had become worse. At the same time, the carrier/noise ratio for the L2 frequency remained unchanged. The flare of the X-3 level (02.40 UTC, December 13, 2006) led to a simultaneous deterioration of the carrier/noise ratio for both frequencies. The duration of the phenomena observed in both cases was about 30 minutes.

The next phenomenon that deserves attention is the increasing power of the signal received by GPS receiver during the period of strong solar activity. The time-dependence of the power of the GPS signal, and an integral number of failures at the receiver during a geomagnetic disturbance on July 15, 2000 was published in [57]. There was an increase of about three times in the intensity of the signal at the receiver with respect to satellite signal power. The integral number of failures grew with increasing intensity of the received signal. The authors did not give an explanation for the growth of intensity.

The observations show that a significant increase of high-frequency (UHF) radiation from the upper atmosphere is a result of solar flares. The intensity of UHF radiation in

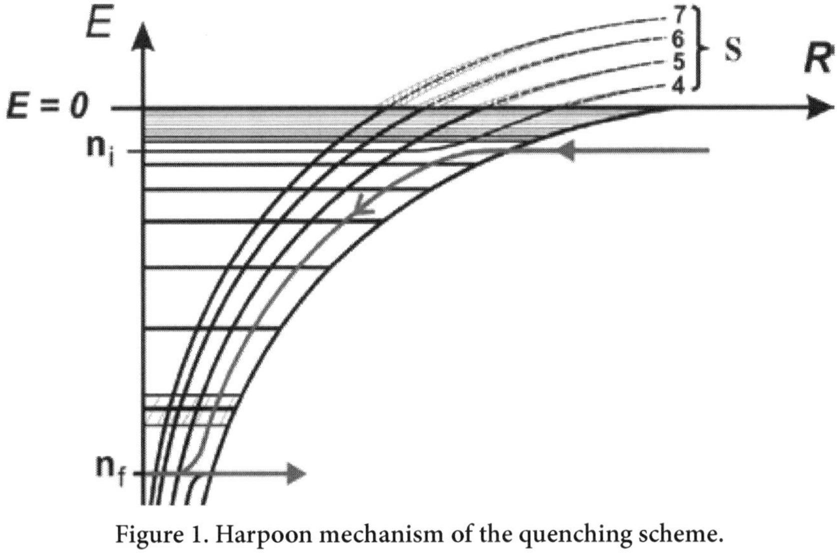

Figure 1. Harpoon mechanism of the quenching scheme.

such events in 40 or more times higher than typical levels of solar microwave bursts [58]. In particular, such events were observed during periods of geomagnetic disturbances, such as observations at wavelengths in the 3-50 cm range [59]. The observations were made simultaneously at several points within the project SETI. The intensities were not given in this paper. The corresponding graphical dependences were normalized to the maximum value. Analysis of different possi- bilities to generate the observed UHF radiation has shown that the largest contribution to the resulting picture of the spectrum is made by transitions between Rydberg levels of neutral components of the non-equilibrium two-temperature plasma. They are excited by the action of solar radiation flux or a stream of electrons emitted from the upper ionosphere in a collision with electrons [55].

Highly excited states of atoms and molecules are called Rydberg states if they are located near the ionization limit and are characterized by the presence of an infinite sequence of energy levels converging to the ionization threshold. They represent an intermediate between the low- lying excited states and ionized states located in the continuous spectrum. Rydberg atoms and molecules have one excited weakly bound electron, the state of which is characterized by the energy level and angular momentum l with respect to the ion core. Energy levels with large angular momenta do not depend on l. They are so-called orbitally degenerate states. These Rydberg states are statistically the most stable when the electron spends most of the time at large distances from the ion core. The process leading to the formation of these states is called l-mixing. In the upper atmosphere l-mixing flows quickly and is irreversible, i.e. almost all Rydberg particles are in degenerate states. As a result, the differences between atoms and molecules are lost, and the range of UHF radiation is homogeneous (i.e. not dependent on the chemical composition of the excited particles) [60]. The l-mixing process takes place in a dense neutral gas medium with a concentration greater than 10^{-12} cm^{-3}, which corresponds to the altitude h ≤ 110 km. The criterion for the efficiency of the process is the condition that the volume of electron cloud of the Rydberg particle A^{**} (with radius $2n^2a_0$, where a_0 is equal to the Bohr radius) has at least one neutral molecule *M*. The interaction between them leads to the formation of quasimolecules $A^{**}M$. The potential energy curves of these molecules split off from degenerate Coulomb levels and are classified according to the value of the angular momentum L of a weakly bound electron with respect to the molecule *M* [60]. The shape of the potential curve is determined by the characteristics of the elastic scattering of slow electrons on molecule M.

The optical transitions between split and degenerate states of quasimolecule $A^{**}M$ that occur without changing the principal quantum numbers ($\Delta n = 0$) correspond to radiation in the decimeter range [61].

Rydberg states of the particles A^{**} are not populated at altitudes h ≤ 50km due to the quenching process. This takes place at the time of interaction of Rydberg particles with unexcited molecules of oxygen during the intermediate stage of ion complex $A^+(nL)O_2^-(s)$ creation (harpoon mechanism), i.e.

$$A^{**}(n_i L_i) + O_2 \rightarrow A^+(nL)O_2^-(s) \rightarrow A^{**}(n_f L_f) + O_2,$$

where s is the vibrational quantum number. It is due to the fact that the negative molecular ion O^{-}_{2} has a series of resonance vibrationally excited autoionizing levels located on the background of the ionization continuum (see Figure 1). Taking into account these two factors, it can be assumed that the layer of the atmosphere emitting in the decimeter range is formed between 50 and 110 km.

The increased solar activity leads to the formation of two types of non-equilibrium plasma in the D- and E-layers of the ionosphere. They are *recombination plasma* and *photoionization plasma*. The first type corresponds to a non-equilibrium two-temperature plasma in which Rydberg states are populated by collisional transitions of free electrons into bound states of the discrete spectrum due to inelastic interaction with the neutral components of the environment [61]. The electron temperature T_e here varies from 1000 to 3000 K [62], and the ambient temperature T_a of this layer varies from 200 to 300 K depending on altitude. Note that in the lower D-layer, the collisional mechanism of Rydberg-state population is dominant. Thermalization of electrons takes place mainly due to the reaction of the vibrational excitation of molecular nitrogen

$$e^{-} + N_2(v=0) \rightarrow N2- \rightarrow e^{-} + N^2(v\geq 1)$$

leading to the formation of an intermediate negative ion.

The second type corresponds to a photoionization plasma in which the population of Rydberg states occurs under the action of the light flux coming from the solar flare. In this case, the quantity of the Rydberg states is determined by the intensity of incident radiation. Here, in contrast to the recombination plasma, the low-lying Rydberg states are also populated, which leads to production of the infrared radiation.

7. The radiation of recombination plasma in the decimeter band

Let us consider the effect of molecules *N2* and *O2* on the UHF spectrum of spontaneous emission (absorption) in the decimeter band which appears in the D- and E-layers of Earth's ionosphere during strong geomagnetic disturbances [63]. Since the concentration of free electrons n_e is small compared with the concentration of atmospheric particles ρ_a both in normal conditions and in case of a magnetic storm, there is no noticeable change in the ambient temperature T_a [64,65]. Both for night and day time, it is of the order of thermal temperature in the D- and E- layers [66]. It is due to the fact that a high translational temperature of the particles coming from the ionosphere (F-layer and above) when entering a denser medium is spent on the vibrational and rotational excitation of atmospheric molecules. Further relaxation of the excitation occurs in the process of resonant transfer of internal energy. Its transport is carried out in subsequent collisions. The vibrational temperature T_v should relax faster than the rotational T_N. The temperatures T_a and T_N usually equalize fairly quickly because of the rapid exchange of energy between rotational and translational motions. The rate of transmission increases with increasing excitation energy of the medium. Thus, in the D-layer the electron temperature Te separates from the ambient temperature T_a, and non-equilibrium quasistationary two-temperature recombination plasma is set, where $T_a << T_e$, which agrees well with direct measurements [66].

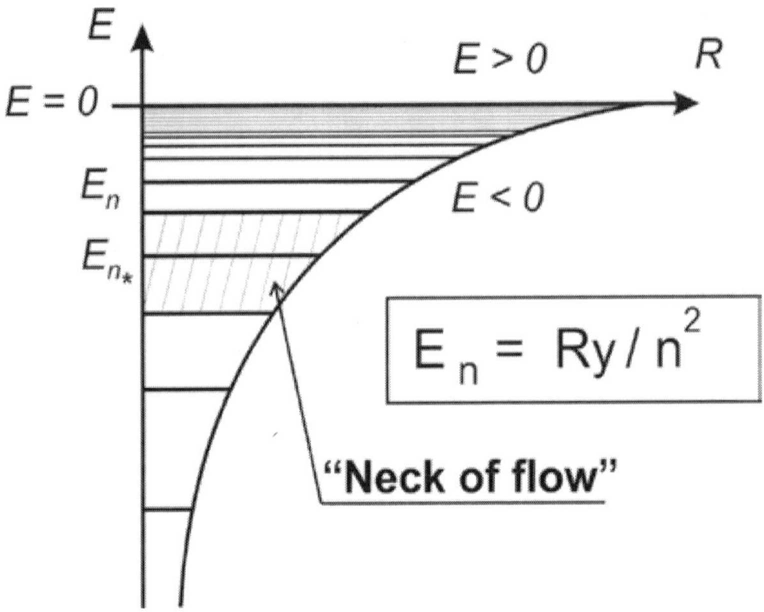

Figure 2. Energy scheme for neck-of-flow location.

Assuming that the electron flow is stationary and their concentration n_e slightly varies over time, the populations of excited states of Rydberg atoms and molecules $m_n(n_e, T_e)$ can be determined from the balanced equations taking into account the processes of recombination, ionization and radiation [67]. Since the frequency of collisions of electrons with medium particles is equal to 10^{12}-10^{14} s^{-1}, two population distributions of atoms and molecules on excitation energies of discrete states are formed in the plasma due to rapid energy exchange between bound and free electrons with medium particles and each other. The first distribution with a temperature Te corresponds to highly excited Rydberg states with binding energies E_n, which is less than the characteristic E_* energy. Rydberg states are not populated close to this energy at the $\Delta E << E^*$ interval that is called the bottleneck of the recombination flux or «neck of flow» (see Fig.2). The second population distribution with the ambient Ta temperature is caused by collisions between particles of the medium and refers to the low-lying electronic states where radiative transitions dominate. Thus, two effective regions of population can be identified in the spectrum: the part with $E_n<E_*$, where the collision processes dominate, and the part $E_n >E_*$, where excitation of the states is due to radiative transitions. The population of the levels located between these regions is negligible since they are effectively quenched because of radiative transitions in the IR, visible, and UV range.

Passage of the electrons through the neck of the flow on the energy scale is the slowest stage of the process that determines the kinetics of non-equilibrium plasma. The energy E_* is found from the condition of minimum of quenching rate constant due to transitions to low-lying states and is defined as [67]:

$$E_* = \frac{Ry}{n_*^2} \approx 27.21 \left[\frac{n_e}{4.5 \cdot 10^{13}} \left(11600 / T_e \right)^2 \right]^{1/4} . \qquad (30)$$

(here concentration of electrons n_e is in cm^{-3} and temperature T in K). The dependences of n_* on the concentration n_e and temperature T_e of free electrons calculated according to (30) are listed in Table 1.

The Rydberg state perturbations by the neutral particles are reduced to the splitting of degenerate levels of the highly excited $A{*}{*}N_2$ and $A{*}{*}O_2$ quasimolecules in groups of sublevels whose positions are determined by the characteristics of weakly bound electron scattering on the nitrogen and oxygen molecules [60]. Moreover, the transitions between the $L = 0 - 3$ terms, split off from the Coulomb levels, provide the greatest contribution to the UHF radiation, where L is the angular momentum of the weakly bound electron with respect to the perturbing particle of the medium [63]. A non-equilibrium two-temperature plasma is formed at $10^{12} < \rho_a < 10^{16}$ cm^{-3} densities when the detachment of the electron temperature T e from the medium temperature T a occurs under the influence of the flux of ionospheric electrons. Populations of Rydberg particles generated here are defined by the temperature T e, concentration n_e and flow of free electrons, and medium concentration ρ_a. Increase in concentration in the lower part of the D-layer leads to an increase in the shock quenching rate of Rydberg states. At $\rho a \geq 10^{16}$ cm^{-3} their population decreases sharply. The spatial region for the formation of excited particles concentrates in the lower part of the E-layer and within the D-layer closer to its upper boundary (this region is 25-30 km wide), where the states with principal quantum numbers $n \approx 20 \div 60$ will be populated most effectively [63].

The states with small momenta $l \leq l^*$ are perturbed strongly by the ionic core (for example, $l^* = 3$ in the nitrogen and oxygen molecules). The perturbing particle M field in turn only influences such superpositions of Coulomb center states that have low electron angular momenta L with respect to M. These two «different center» groups of terms will be called *Rydberg l* and *degenerate L* terms. The latter are defined as [60]

$$U_n^{(LL_Z)}(R) = -\frac{1}{2\left[n - \mu_{LL_Z}(R)\right]^2} - \frac{\beta}{2R^4}, \tag{31}$$

where quantum defects induced by the field of the molecule M are equal to

$$\mu_{LL_Z}(R) = -\frac{1}{\pi}\arctan\left[\kappa(R)K_{L,L_Z}(R)\right],$$
$$K_{L,L_Z}(R) = a\delta_{L0} - \frac{L(L+1)-3L_Z^2}{L(L+1)(2L-1)(2L+3)}Q(1-\delta_{L0}) - \frac{\pi\kappa(R)}{(2L-1)(2L+1)(2L+3)}\left[\beta + \frac{L(L-1)-3L_Z^2}{(2L-1)(2L+3)}\beta'\right]. \tag{32}$$

Table 1. Dependency of neck-of-the-flow n_* position on the concentration n_e and temperature of free electrons T_e.

Concentration of free electrons n_e, см$^{-3}$	Effective neck of the flow principal quantum number n^* at various T_e values (K)		
	T_e = 1000K	T_e = 2000K	T_e = 3000K
10^1	77	91	101
10^2	58	69	76

10^3	43	52	57
10^4	33	39	43
10^5	25	29	32
10^6	19	22	24
10^7	14	17	18
10^8	11	13	14
10^9	8	10	11
10^{10}	6	7	8
10^{11}	5	6	6
10^{12}	4	4	5

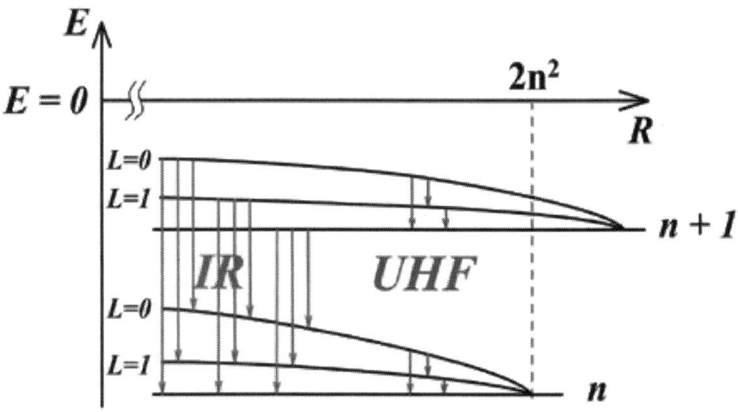

Figure 3. Potential energy curves of the quasimolecule *AM*.**

Here a is the length of electron scattering on the perturbing particle M, Q is its quadrupole moment, β and β' are the isotropic and anisotropic polarizability components, κ (R) is the semiclassical momentum of a weakly bound electron in the Coulomb field, R is the interatomic distance of the A ** M quasimolecule, quantity $\nu = [-\ 2E\ ]^{1/2}$, where E is the total energy of the Rydberg particle measured from the ground state of ion A + (here and throughout, we use the atomic system of units with $h = e = m_e = 1$).

Figure 3 shows the energy diagram of the potential energy curves of quantum L-states depending on interatomic coordinate R. For angular momenta $L=0$ and $L=1$ they are split from degenerate Coulomb levels with principal quantum numbers n+1 and n in the classical turning points. Blue arrows indicate the optical transitions between the split and degenerate states of the quasimolecule $A^{**}M$ occurring without changing the principal quantum number $\Delta\ n = 0$. This corresponds to UHF radiation in the decimeter range. The red arrows show similar transitions with a change in principal quantum number $\Delta\ n = 1$ which correspond to IR radiation.

Radiation intensities of photon energy emission per unit time for $L \rightarrow L'$ and $L \rightarrow n$ without changing principal quantum number $\Delta\ n = 0$ defined as

$$I_{LL'} = \overline{W_n(L \to L')} = \frac{16\,\omega^4(L \to L')}{9c^3} C^2_{v_L}\ LL',$$

(33)

$$I_{Ln} = \overline{W_n(L \to n)} = \frac{16\,\omega^4(L \to n)}{9c^3} \sum_{l \geq l^*}^{n-1} C^2_{v_L}\ Ll',$$

(34)

where $\omega\ (L \to L')$ and $\omega\ (L \to n)$ are the appropriate transition frequencies, $v_L\ (n) = 1 / \sqrt{-2U_n^{(L,0)}\,(R = n^2)}$ is the effective principal quantum number of L term. The procedure for calculating the coefficients C_{vL} L l and C_{vL} L l is described in detail in [63]. Averaging of the intensities (33) and (34) is over all possible interatomic coordinates R as follows

$$\overline{W_{nL}} = \int W_{nL}\ (R)\,\Psi^2_{nL}\ (\bar{r} = n^2,\ \mathbf{R})\ d\,\mathbf{R}$$

($\Psi\ n\ L$ is the electron wave function of $A^{**}M$ quasimolecule). This corresponds to a static consideration of the problem, which is valid if the average electron velocity $1 / n$ is much higher than the characteristic velocity of atomic particles in the medium when we have the inequality $n\sqrt{2T_a/M_a} << 1$ (where Ma is the reduced mass). Under our conditions, this situation is certainly satisfied. As the main distortion of the Coulomb wave function is in the vicinity of the perturbing particle M, we can put to the first approximation that $\overline{W_{nL}} \approx W\,n\,L\,(R = n^2)$. Considering also that radiation from the states with a given value of v_L is not coherent, we should add consistently the transition intensities (33) and (34) for the corresponding emission lines of the $A^{*}{}^{*}N_2$ and $A^{*}{}^{*}O_2$ quasimolecules separately in or- der to determine a total contribution of all possible transitions and multiply them by the ef- fective population of the Rydberg L term $m_{eff}\,(n, \omega, T_e) = P_n\,(\omega)\,m_n\,(n_e, T_e)$. By analogy with [67] they are calculated as

$$m_{\text{eff}}(n,\omega,T_e) = P_n(\omega)\,\frac{n_e^2(T_e)}{\Sigma_i}\left[\frac{2\pi}{T_e}\right]^{3/2} n^2 \exp(1/2n^2T_e)$$

(35)

(Σ_i is the statistical sum of positive molecular ions). The non-equilibrium plasma factor P n (ω) taking into account the flux of precipitating electrons from the ionosphere and the quenching of Rydberg states of the particles can be represented as

$$P_n(\omega) = F_n(T_e,\omega)\exp\left[-f_n\left(\rho_a, T_a\right)\right].$$

(36)

The coefficient F_n is given by the flow of electrons, and the value of $f_n\,(\rho_a, T_a) = 16\,\rho a\,n^2 / T_a$ characterizes the decrease in the populations of m_{eff} due to quenching of the Rydberg particles in a neutral medium [63]. It is defined as the ratio of the effective number n_{eff} of electrons, moving in a flow with velocity $\sqrt{2T_e}$ and passing the distance $\sim\sqrt{2T_e}\,\tau_{eff}$ through a unit area during the lifetime τ_{eff} of the split-off *L* excited state, to the equilibrium concentration n e of electrons, i.e.

$$F_n(T_e,\omega)=n_{\mathrm{eff}}(T_e,\omega)/n_e=\sqrt{2T_e}\,\tau_{\mathrm{eff}}(\omega), \tag{37}$$

The lifetime of degenerate Rydberg states of the quasimolecule characterized by the time of spontaneous emission in the infrared (i.e. $\tau_{eff} \simeq \tau_R$) which is defined as $\tau_R^{-1}(\omega) = W_n(\Delta n = 1)/n\omega$ where $W_n(\Delta n = 1)$ is the intensity of the energy emission in IR region. Since with the frequency increase the radiation lifetime decreases as $\tau_R \sim \omega^{-5/3}$ [68], the gain factor can be expressed in the form

$$F_n(T_e,\omega)\simeq\sqrt{T_e/T_{\max}}\,(n/n_{\max})^5F_{\max}=\sqrt{T_e/T_{\max}}\,(\omega_{\min}/\omega)^{5/3}F_{\max}, \tag{38}$$

where the maximum is achieved at a frequency ω = n – 3 and electron temperature $T_{max} = 3500$ K. Under these conditions, the value of $F_{max} \sim 10^{10}$. Quantity n_{max} is the position of the maximum population of Rydberg states, which is defined as [63]

$$n_{\max}=\left[\frac{2^{17}\sqrt{2}\pi^7}{675}T_e^{3/2}\rho_a^2\Big(\sum_i\alpha(i)B_iQ_i^2\Big)\Big(\sum_{i'}\alpha(i')Q_{i'}^2\Big)\right]^{-1/15}. \tag{39}$$

Here α (*i*) denotes the weight factors of N_2 and O_2 molecules equal to $\alpha(N_2) \simeq 0.79$ and $\alpha(O_2) \approx 0.21$ respectively. The values of Q_i and B_i are their quadrupole moments and rotational constants expressed in $e\,a_0^2$ units. It will be recalled that in the system of the atomic units, the concentration ρ_a of medium, rotational constant B_a, and medium temperature T_a are determined as

$$\rho_a(\text{a.u.}) \simeq (0.52917)^3\cdot 10^{x-24},\ B_a(\text{a.u.}) = B_a(\text{cm}^{-1})/8066\cdot 27.21,\ T_a(\text{a.u.}) = T_a(K)/11600\cdot 27.21$$

(*x* is the exponent that characterizes the medium concentration which in these conditions varies from 12 to 16). Note, that according to (10) the position of the Rydberg state population maximum $m_{max}(T_e, \rho_a)$ along the n axes decreases as $\sim T_e^{-1/10}$ and $\sim \rho^{-2/15}$ as the temperature of electrons and medium concentration increase. In the calculations performed below, the spectroscopic parameters of molecules and molecular ions of nitrogen and oxygen are taken from [69]. Scattering lengths, quadrupole moments, and static polarizabilities of the nitrogen and oxygen molecules, for which data from [70-74] were used, are given in Table 2. The corresponding dependences are shown in Table 3.

Table 2. Scattering lengths a, quadrupole moments Q, and static polarizabilities β of oxygen and nitrogen molecules *)

Molecules	a	Q	β
O_2	1.60 [70]	– 1.04 [72]	10.6 [73]
N_2	0.75 [71]	– 1.30 [72]	11.1 [74]

**) Because the contribution of the second term in square brackets in (3) is small, we do not give β' values for these molecules*

The intensity of the UHF radiation per unit volume is written as follows:

$$W_{tot}(\omega) = \sum_{i=1,2} \alpha(i) \sum_{nS} m_{nL}(i) \overline{W_n^{(i)}(L \to S)}. \tag{40}$$

The S index in (40) takes on various values ($L' > L$ or $L' = n$) according to the sequence in which radiation lines intersect the straight-line ω frequency. The distribution of n-dependent emission lines (corresponding to $L = 0 - 3$ values) for N_2 and O_2 molecules contain the series of four $L \to L'$ transitions which converge with increasing L' to the limiting $L \to n$ transition [63].

The shift in frequency of these limits for $N_{L\,n}$ and $O_{L\,n}$ series leads to the heterogeneity of the UHF spectrum, in which there are three spectral ranges of frequencies. The first includes the frequencies ω = 1.17 – 1.706 GHz, the second corresponds to the $\omega = 4.31 - 6.094$ GHz range, and third contains the $\omega = 7.27 - 10.00$ GHz one.

The optical thickness of the radiating layer $H_R(\omega)$ is defined by the limiting value of the medium concentration $\rho_a(H_R(\omega))$ from which the value of $W_n^{(i)} = \sum_S W_n^{(i)}(L \to S)$ vanishes. This situation occurs when the concentration of electrons $n_e \leq 10^6$ cm^{-3} for medium concentration $\rho_a = 10^{15}$ cm^{-3}. On the other hand, the dependence of m_{nL} $(\rho_a(z))$ is determined by the quenching process of Rydberg particles and drops sharply when $\rho_a \geq 10_{16}$ cm^{-3} [63]. In these circumstances, a quenching is accompanied by the transfer of electronic excitation energy into translational motion of the heavy medium components. Since the concentration of excited particles is small compared with the medium concentration, the process of cooling flow of free electrons in the rotational

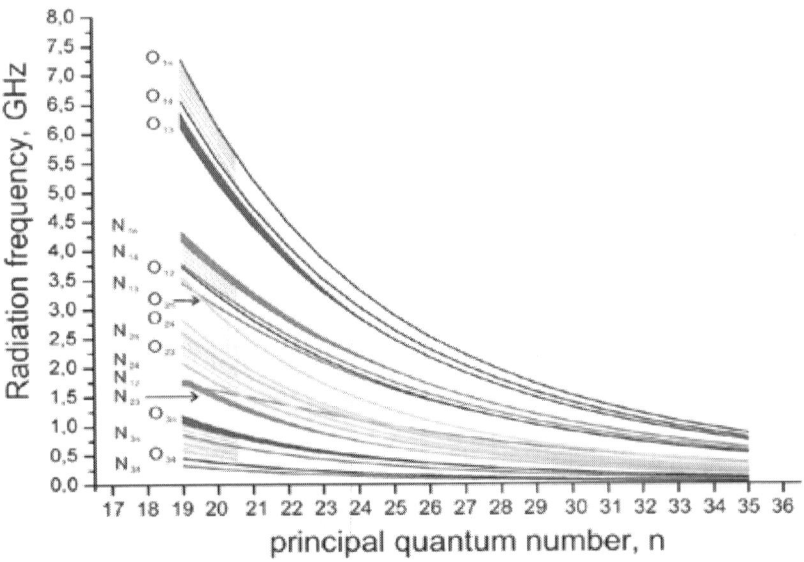

Figure 4. UHF radiation lines of the A**N2 and A**O2 quasimolecules.

transitions of molecules should be carried efficiently in the bottom of the D-layer. The Rydberg states are eliminated when the Te and T a temperatures are equalized. This condition is the criterion for the lower boundary Hp of the plasma layer, which does not depend on the radiation frequency and corresponds to $\rho_a \sim 10^{16}$ cm^{-3}.

Figure 4 shows the dependence of the frequency of the emission lines for $A^{**}N_2$ and $A^{**}O_2$ quasimolecules on the principal quantum number n for the transitions $L \to n$ between the split-off and degenerate Coulomb levels and the transitions $L \to L'$ between the split-off levels, which are calculated according to formulas (31) and (32). At the 0.8 - 10 GHz frequency range the distributions with respect to L for the radiation lines depending on n for each quasimolecule contain the series of four lines, corresponding to the $L \to L'$ transitions, which for increasing L' converge in the limit to the L → n transitions, where L = 0, 1, 2, 3. These lines are marked with symbols $N_{LL'}$, N_{Ln} and $O_{LL'}$, O_{Ln} for $A^{**}N_2$ and A**O2 quasimole-cules, respectively.

Table 3. Dependency of the position of maximum population n_{max} of Rydberg states on medium concentration ρ_a and temperature T_e of free electrons

ρ_a, cm^{-3}	n_{max}			
	T_e=1000K	T_e=2000K	T_e=3000K	T_e=4000K
10_{12}	79	74	71	69
10_{13}	58	54	52	50
10_{14}	42	40	38	37
10_{15}	31	29	28	27
10_{16}	23	21	20	20

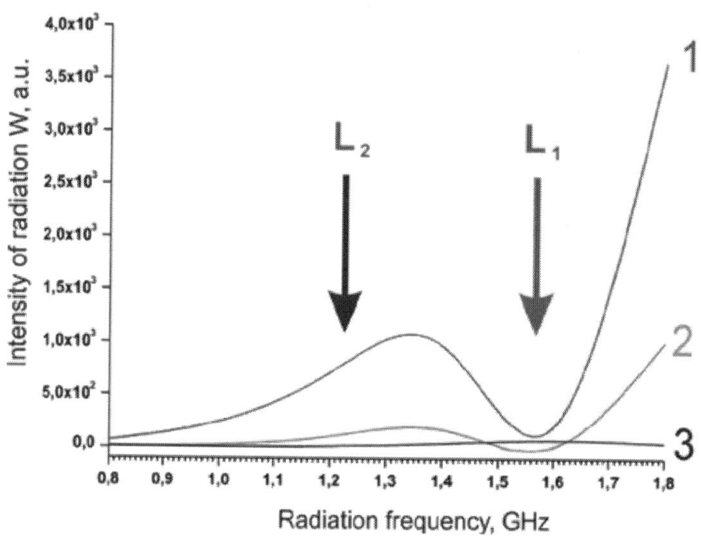

Figure 5. The spectrum of UHF radiation for a quiet ionosphere. Curve (1) corresponds to electron temperature T =1000 K and medium concentration $\rho a = 10^{12} - 10^{13}$ cm^{-3}; (2) T_e =1000 K, $\rho_a = 10^{14}$ cm^{-3}; (3) T_e =2000 K, $\rho_a = 10^{12} - 10^{13}$ cm^{-3}

It is seen that the frequency shift for the NL n and OL n limits relative to each other leads to the formation of three spectral ranges, which transitions are suppressed for small values of n. This difference is caused by the characteristics of slow-electron scattering by the nitrogen and oxygen molecules.

The intensities of incoherent UHF radiation for the excited medium calculated by the formula (11) in the 0.8-1.8 GHz range are presented in Fig.5 and Fig.6 as functions of frequency for quiet and disturbed ionosphere states. It is shown that the profile of UHF radiation is a non- monotonic function of the radiation frequency and increases sharply near the right edge of the range.

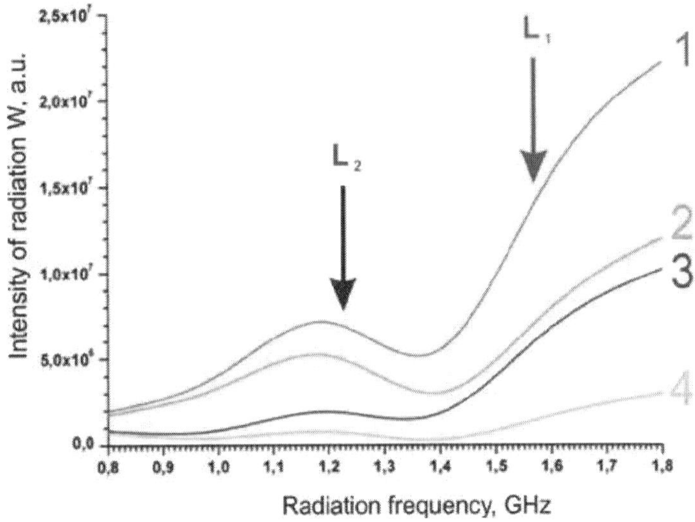

Figure 6. The spectrum of UHF radiation for a disturbed ionosphere. Curve (1) corresponds to electron temperature T_e =2000 K, and neutral medium concentration $\rho_a = 10^{12} - 10^{13}$ cm^{-3}; (2) T_e =2000 K, $\rho_a = 10^{14}$ cm^{-3}; (3) T =3000 K, $\rho_a = 10^{12} - 10^{13}$ cm^{-3}; (4) T_e = 3000 K, $\rho_a = 10_{14}$ cm^{-3}.

With increase in electron concentration n_e by two orders of magnitude, the relative intensity W increases by about four orders of magnitude. This is likely directly related to the observed effect of sequential disappearance of the L_1 and L_2 frequencies of GPS signal as the power of the sun storm increases [56] because the position of the first radiation intensity attenuation region (1.17 – 1.71 GHz) and the «transparency window» of the propagation of satellite signals nearly coincide.

8. Radiation of photoionization plasma

The photoionization plasma is formed during 20-30 minutes under the influence of wideband radiation coming into the atmosphere after a solar flare has occurred. This process is caused by multi-quantum excitation of the electronic states of nitrogen and oxygen atoms and molecules. In this case the spin-forbidden character of the corresponding radiative transitions is removed due to interaction of excited particles with ambient molecules M and creation of the $A^{**}N_2$ and $A^{**}O_2$ quasimolecules. The distribution of

Rydberg-state populations for large values of principal quantum number $n > 30$ is similar to that discussed above in non-equilibrium recombination plasma. A difference lies in the fact that the top of the energy scale is further depleted by photoionization of the Rydberg states. For small values n < 10, the process of quenching of Rydberg and low-lying excited states takes place because of predissociation including non-adiabatic transitions through intermediate valence configurations as well as the resonant (non-resonant) transfer of internal energy due to collisional processes with subsequent thermalization of the medium. The effect mentioned above is confirmed by an observed increase in temperature of the neutral environment Ta with increasing height in the range 40-60 km. Thus, in photoionization plasma the orbitally degenerate L-states are populated primarily for principal quantum numbers 10<n<30. In contrast to the recombination plasma, the popu- lation-distribution function here is dependent on the intensity of light and is not a function of either the temperature of free electrons Te or the resulting shape of the absorption line.

Under these conditions, spontaneous decimeter radiation (which fits to $\Delta\ n = 0$) will be accompanied by strong IR radiation (with $\Delta\ n = 1$), the intensity of which exceeds by at least three orders of magnitude the intense UHF radiation of recombination plasma [61]. This can be explained by the fact that according to [63], the maximum of IR radiation for the recombi- nation plasma accounts for the interval of principal quantum numbers $20 < n < 30$, and in case of the photo-ionization plasma, the maximum should be located in the vicinity of $n = 10$. It should also be noted that the l-mixing process for the low-lying Rydberg states is suppressed, and the influence of the environment should be significantly reduced [60]. This is particularly important for the formation of the frequency spectrum profile of the IR radiation under conditions of strong solar activity. The maximum of the radiation should occur near the bottom of the D-layer at an altitude of 50-60 km. This is due to the fast decrease ~ n^4 of the l-mixing cross section with increasing principal quantum number n.

Table 4. Dependence of principal quantum number $\bar{n}$ on medium concentration ρ_a

ρ_a, 10^{12} cm_{-3}	1	10	10^2	10^3	10^4	10^5
n	77	52	36	24	16	11

This phenomenon should lead to the population emptying of the low-lying molecular Rydberg states as a result of predissociation [68]. In reality, it is necessary that l-mixing occurs, i.e. inside the Rydberg-electron cloud of the A** particle, at least one medium molecule M should react to form an $A^{**}M$ quasimolecule. This condition corresponds to the principal quantum number

$$\bar{n} = 10^4 \left[\frac{3}{32\pi\,(0.53)^3\,\rho_a} \right]^{1/6}, \tag{41}$$

which depends on the concentration of the medium ρ_a. Table 4 shows that it is realized at a concentration of about 10^{17} cm^{-3}. This means that the process of l-mixing plays an important role in shaping the frequency profile. No less important here are the processes of quenching of Rydberg states, which should reproduce the simple dependence of intensity I on n at the left long-wave side of the profile, for example, as $n^2 \, exp\,(-\alpha n^2)$ [63]. The slope of this curve depends on the ratio of luminous flux and the rate of quenching.

9. Conclusion

The success of the GPS methodology rests on an objective theory of measurement which denies Einstein's symmetry principle derived on the basis of the LT. There is never any ambiguity in principle as to which of two clocks in relative motion runs slower. The goal of relativity theory is to quantitatively predict relative rates of clocks on the basis of information regarding their respective states of motion and positions in a gravitational field. For this purpose it is necessary to designate a specific rest frame (ORS) to act as reference in determining the speeds of the clocks to be used to compute the amount of time dilation. The Earth's center of mass (ECM) serves as the ORS for computing the rates of clocks located on orbiting satellites as well as on the Earth's surface. The effects of gravity on the relative rates can then also be computed from knowledge of their respective positions in space relative to the ECM. To the surprise of many, it was found that the transverse Doppler effect was asymmetric when the light source and detector are mounted on a high-speed rotor, thus agreeing with Einstein's alternative inter- pretation of time dilation that rejects the above symmetry principle and concludes instead that an accelerated clock always runs slower than its stationary counterpart. The EP was used to obtain a quantitative prediction of these results, but it was pointed out by Sherwin that the lack of symmetry expected from the LT *proves that it is only valid for uniform translation.* This conclusion has been largely ignored, however, and emphasis has been placed instead on the possibility of using the experimental data to quantitatively predict the dependence of the rates of clocks on their state of motion and position in a gravitational field. While this information has proven invaluable in the development of GPS, it leaves open the question of whether other predictions of the LT such as space-time mixing, remote non-simultaneity, Fitzgerald-Lorentz length contraction (FLC) and Lorentz invariance are valid under similar circumstances in which acceleration is present, and even whether the LT has validity for uniformly translating systems.

In considering this point it has been noted that Einstein made an undeclared assumption, a *hidden postulate*, in his derivation of the LT which he made no attempt to justify. An example has been given to show that this assumption in fact leads to contradictory predictions regarding whether two observers in uniform relative motion agree or disagree on the length of an object that is oriented perpendicularly to their velocity (clock riddle). It has thereupon been shown that all hitherto successful predictions of relativity theory, including those that are relevant to GPS, can be obtained by assuming that clock rates of observers have a constant ratio as long as neither their relative state of motion nor their respective positions in a gravitational field change. The latter assumption leads to an alternative Lorentz transformation (ALT) which is compatible with Einstein's two postulates of relativity and the same velocity transformation (VT) that is obtained using conventional STR. This version of the theory eliminates the contradiction of the clock riddle while continuing to successfully predict the aberration of starlight at the zenith and the results of the Fresnel light-drag experiment. The resulting theory recognizes that the rates of proper clocks do differ between inertial systems, but that it is nonetheless impossible for a given observer to determine his state of motion on the basis of strictly in situ measurements. This is because there is a uniform scaling of the rates of all natural clocks when they are accelerated by the same amount. The same holds true for all other physical quantities. This conclusion is perfectly consistent with the RP, which can be restated as follows: *The laws of physics are the same in all inertial systems but the units in which they are expressed can and do differ in a systematic manner.*

The difference between the impacts of photoionization and recombination plasma on the distortion of the GPS satellite signals is clearly manifested in the observed dependence of the positioning errors with respect to time. In the first case, there is sharp and narrow (up to 20 min) peak with a positioning error of more than 50 m. The second case corresponds to the formation of a bell-shaped dependence of the error with a typical width of several hours and positioning errors of 15-20 m.

During periods of solar activity, characteristic IR radiation should be formed, which is indicative of Rydberg states having been formed in the D-layer of the Earth's ionosphere. The integrated intensity close to the IR radiation maximum gives information about the population of Rydberg states at altitudes of 50-60 km. These results can be used as a starting point for the corresponding kinetic calculations. The slope on the left side of the frequency profile contains information about the magnitude of the light flux.

An independent line of the research may be a systematic analysis of the long-wave IR spectrum of radiation (for $\Delta n = 1$ transitions) emitted by the D-layer for a long time during a strong geomagnetic storm. This band of the spectrum falls in the range of principal quantum numbers $20 < n < 40$, where the l-mixing process is very efficient. The features of the frequency profile described above should also appear in this case. By contrast, the spectral appearance of $n\ L \rightarrow (n-1)\ L'$ transitions will be richer with the radiation of $A^{**}N^2$ and $A^{**}O^2$ quasimolecules in the decimeter range (for the $\Delta n = 0$ transitions). This effect may serve as a plasma diagnostic and IR-tomography of the ionosphere D-layer.

In conclusion, it should be noted that two physical factors play key roles during the propaga- tion of GPS satellite signals. The first is the absorption and subsequent stimulated emission of electromagnetic waves on Rydberg states of the $A^{**}N_2$ and $A^{**}O_2$ quasimolecules with a time delay of 10^{-5}-10^{-6} seconds in a single scattering process. The second is due to incoherent plasma radiation. These processes are applied independently of each other. The increase in two to three times in the envelope of the resonance intensity profile and the formation of a phase shift are the most characteristic features for the resonance absorption of electromagnetic waves with subsequent re-emission. These two processes alone can explain the power growth and the disappearance of GPS signal observed in [57]. This means that the frequencies of the signals emitted by the satellite and subsequently received on the ground may differ substantially from one other.

Thus, the physical cause of the time delay and phase shift of the GPS satellite signal during periods of strong solar activity is associated with a cascade of resonant re-emission of Rydberg states of the A**N2 and A**O2 quasimolecules in the E-and D-layers of the ionosphere, i.e. they are determined by quantum properties of the propagating medium.

References

1. Einstein A. Zur Elektrodynamik bewegter Körper. Annalen der Physik 1905;322(10) 891-921.
2. Wilkie T. Time to remeasure the metre. New Scientist 1983;27 Oct. 258-263.
3. Will C M. Was Einstein Right?: Putting General Relativity To The Test. 2nd ed. New York: Basic Books Inc; 1993, p. 272.
4. Voigt W. Ueber das Doppler'sche Princip. Göttinger Nachrichten 1887;7 41-51.
5. Pais A. 'Subtle is the Lord ...' The Science and the Life of Albert Einstein. Oxford University Press; 1982, p. 125.
6. Lorentz H A. Electromagnetic phenomena in a system moving with any velocity smaller than that of light. Proceedings of the Royal Netherlands Academy of Arts and Sciences 1904;10 809-831.
7. Poincaré H. La mesure du temps. Revue de métaphysique et de morale 1898;6 1-13.
8. Larmor J. Aether and Matter. Cambridge University Press; 1900.
9. Fitzgerald G F. The Ether and the Earth's Atmosphere Science 1889;13(328) 390.
10. Lorentz H. A. The Relative Motion of the Earth and the Ether. Zittingsverlag Akad. V. Wet. 1892;1 74–79.
11. Pais A. 'Subtle is the Lord ...' The Science and the Life of Albert Einstein. Oxford University Press; 1982, p. 137.
12. Dingle H. Science at the Crossroads. London: Martin, Brian & O'Keefe; 1972.
13. McCausland I. Problems in Special Relativity. Wireless World 1983;Oct. 63-65.
14. Phipps Jr. T E. Old Physics for New - A Worldview Alternative to Einstein's Relativi- ty Theory. Montreal: Apeiron; 2006, p. 142.
15. Goldstein H. Classical Mechanics. Reading, Massachusetts: Addison-Wesley Publishing Co.; 1950, p. 193.
16. v. Laue M. Die Relativitätstheorie. vol. 1, 7th ed.. Braunschweig: Vieweg; 1955, p. 36.
17. Einstein A. Relativitätsprinzip und die aus demselben gezogenen Folgerungen. Jahrbuch der Radioaktivität und Elektronik 1907;4 411-462.
18. Ives H E, Stilwell G R. An Experimental Study of the Rate of a Moving Atomic Clock. Journal of the Optical Society of America 1938;287 215-219; contd. in: 1941;31(5) 369-374.
19. Mandelberg H I, Witten L. Experimental Verification of the Relativistic Doppler Effect. Journal of the Optical Society of America 1962;52(5) 529-535.
20. Rossi B, Greisen K, Stearns J C, Froman D K, Koontz P. G. Further Measurements of the Mesotron Lifetime. Physical Review 1942;61(11-12) 675-679.
21. Ayres D S, Caldwell D O, Greenberg A J, Kenney R W, Kurz R J, Stearns B F. Comparison of π+ and π- Lifetimes. Physical Review 1967;157(5) 1288-1292.
22. Weidner R T, Sells R L. Elementary Modern Physics. Boston: Allyn and Bacon; 1962, p. 410.

23. Serway R A, Beichner R J. Physics for Scientists and Engineers. 5th ed. Orlando: Harcourt; 1999, p. 1262.

24. Sard R D. Relativistic Mechanics. New York: W. A. Benjamin; 1970, p. 124.

25. Hay H J, Schiffer J P, Cranshaw T E, Egelstaff P A. Measurement of the Red Shift in an Accelerated System Using the Mössbauer Effect in Fe57. Physical Review Letters 1960;4(4) 165-166.

26. Will C M. Clock synchronization and isotropy of the one-way speed of light. Physical Review D 1992;45 403-411.

27. Mansouri R, Sexl R U. A test theory of special relativity: III. Second-order tests. General Relativity and Gravitation 1977;8(10) 809-814.

28. Kuendig W. Measurement of the Transverse Doppler Effect in an Accelerated Sys- tem. Physical Review 1963;129(6) 2371-2375.

29. Champeney D C, Isaak G R, Khan A M. Measurement of Relativistic Time Dilatation using the Mössbauer Effect. Nature 1963;198 1186-1187.

30. Sard R D. Relativistic Mechanics. New York: W. A. Benjamin; 1970, p. 319.

31. Sherwin C W. Some Recent Experimental Tests of the 'Clock Paradox'. Physical Review 1960;120(1) 17-21.

32. Rindler W. in Essential Relativity. New York: Springer Verlag; 1977, p. 32.

33. Buenker R J. The Sign of the Doppler Shift in Ultracentrifuge Experiments. Apeiron 2012;19(3) 218-237. http://redshift.vif.com/JournalFiles/V19NO3PDF/V19N3BU1.pdf

34. Einstein A. Über den Einfluß der Schwerkraft auf die Ausbreitung des Lichtes. Annalen der Physik 1911;340(10) 898-908.

35. Pound R V, Snider J L. Effect of Gravity on Gamma Radiation. Physical Review 1965; 140(3B) 788-803.

36. Einstein A. Die Feldgleichungen der Gravitation. Königlich Preussische Akademie der Wissenschaften 1915; 831–847.

37. Pais A. 'Subtle is the Lord ...' The Science and the Life of Albert Einstein. Oxford University Press; 1982, p. 194.

38. Rindler W. in Essential Relativity. New York: Springer Verlag; 1977, p. 21.

39. Shapiro I I. Fourth Test of General Relativity. Physical Review Letters 1964;13(26) 789-791.

40. Pais A. 'Subtle is the Lord ...' The Science and the Life of Albert Einstein. Oxford University Press; 1982, p. 178.

41. Buenker R J. Use of Hamilton's Canonical Equations to Modify Newton's Corpuscular Theory of Light. Russian Journal of Chemical Physics 2003;22(10) 124-128. http://arxiv.org/ftp/physics/papers/0411/0411110.pdf

42. Buenker R J. Gravitational and Kinetic Scaling of Units. Apeiron 2008;15(4) 382-413. http://redshift.vif.com/JournalFiles/V15NO4PDF/V15N4BU1.pdf

43. Hafele J C, Keating R E. Around-the-World Atomic Clocks: Predicted Relativistic Time Gains. Science 1972;177 166-168.
44. Hafele J C, Keating R E. Around-the-World Atomic Clocks: Observed Relativistic Time Gains. Science 1972;177 168-170.
45. Vessot R F C, Levine M W. A Test of the Equivalence Principle Using a Space-Borne Clock. General Relativity and Gravitation 1979;10(3) 181-204.
46. Buenker R J. Time Dilation and the Concept of an Objective Rest System. Apeiron 2010;17(2) 99-125. http://redshift.vif.com/JournalFiles/V17NO2PDF/V17N2BUE.pdf
47. Van Flandern T. What the Global Positioning System Tells Us about Relativity. In: Selleri F. (ed.) Open Questions in Relativistic Physics. Montreal: Apeiron; 1998. p81-90.
48. Buenker R J. The Global Positioning System and the Lorentz Transformation. Apeir- on 2008;15(3) 254-269. http://redshift.vif.com/JournalFiles/V15NO3PDF/ V15N3BU1.pdf
49. Buenker R J. Simultaneity and the Constancy of the Speed of Light: Normalization of Space-time Vectors in the Lorentz Transformation. Apeiron 2009;16(1) 96-146.
50. Buenker R J. Einstein's Hidden Postulate. Apeiron 2012;19(3) 282-301. http:// redshift.vif.com/JournalFiles/V19NO3PDF/V19N3BU2.pdf
51. Buenker R J. The Clock Riddle: The Failure of Einstein's Lorentz Transformation. Apeiron 2012;19(1) 84-95. http://redshift.vif.com/JournalFiles/V19NO1PDF/V19N1BUE.pdf
52. Langley R B. GPS, the Ionosphere, and the Solar Maximum. GPS World 2000; 11(7) 44-49
53. Markgraf M. Phoenix GPS Tracking System. Flight Report; 2005, VSB 30-DLR- RP-0001
54. http://gps.ece.cornell.edu/realtime.php
55. Golubkov G V, Manzhelii M I, Karpov I V. Chemical Physics of the Upper Atmosphere. Russian Journal of Physical Chemistry B 2011; 5(3) 406-411, DOI: 10.1134/S1990793111030055
56. Cerruti A P, Kintner P M Jr, Gary D E, Manucci A J, Meyer R F, Doherty P, and Coster A J. Effect of Intense December 2006 Solar Radio Bursts on GPS Receivers. Space Weather 2008; 6(10) S10D07, DOI: 10.1029/2007SW000375
57. Afraimovich E L, Astafieva E I, Berngardt O I, Lesyuta O S, Demyanov V V, Kondakova T N, Shpynev B G. Mid-Latitude Amplitude Scintillation of GPS Signals and GPS Performance Slips at the Auroral Oval Boundary. Radiophysics and Quantum Electronics 2004; 47(7) 453-468, DOI: 10.1023/B:RAQE.0000047237.67771.bc
58. Avakyan S V. Physics of the Solar-Terrestrial Coupling: Results, Problems, and New Approaches. Geomagnetism and Aeronomy 2008; 48(4) 417-424, DOI: 10.1134/S0016793208040014

59. Troitskii V S, Bondar' L N, Starodubtsev A M. The Search for Sporadic Radio Emis- sion from Space. Soviet Physics Uspekhi 1975; 17(4) 607–609, DOI: 10.1070/PU1975v017n04ABEH004633

60. Golubkov G V, Golubkov M G, Ivanov G K. Rydberg States of Atoms and Molecules in a Field of Neutral Particles. In: Bychkov V L, Golubkov G V, Nikitin A I (ed.) The Atmosphere and Ionosphere: Dynamics, Processes and Monitoring. New York: Springer; 2010. p1-67, DOI: 10.1007/978-90-481-3212-6_1

61. Golubkov G V. Influence of the Medium on the Electromagnetic Radiation Spectrum of Highly Excited Atoms and Molecules. Russian Journal of Physical Chemistry B 2011; 5(6) 925-930, DOI: 10.1134/S1990793111060108

62. Oyama K I, Abe T, Mori H, and Liu J Y. Electron Temperature in Night Time Spora- dic E-Layer at Mid-Latitude. Annales Geophysicae 2008; 26 (3) 533-541, DOI: 10.5194/angeo-26-533-2008

63. Golubkov G V, Golubkov M G, Manzhelii M I. Microwave Radiation in the Upper Atmosphere of the Earth during Strong Geomagnetic Disturbances. Russian Journal of Physical Chemistry B 2012; 6(1) 112-127, DOI: 10.1134/S1990793112010186

64. Pavelyev A, Tsuda T, Igarashi K, Liou Y A, Hocke K. Wave Structure in the Ionospheric D – and E – layers Observed by Ratio Holography Analysis of the GPS/MET Radio Occultation Data. Journal of Atmospheric and Solar-Terrestrial Physics 2003; 65(1) 59-70, DOI: 10.1016/S1364-6826(02)00226-2

65. Jacobsen K S, Pedersen A, Moen J I, and Bekkeng T A. A New Langmuir Probe Concept for Rapid Sampling of Space Plasma Electron Density. Measurement Science and Technology 2010; 21(8) 085902, DOI:10.1088/0957-0233/21/8/085902

66. Kurihara J, Oyama S, Nozawa S, Tsuda T T, Fujii R, Ogawa Y, Miyaoka H, Iwagami N, Abe T, Oyama K I, Kosch M J, Aruliah A, Griffin E, Kauristie K. Temperature En- hancements and Vertical Winds in the Lower Thermosphere Associated with Auro- ral Heating During the DELTA Campaign. Journal of Geophysical Research 2009; 114(12) A12306, DOI: 10.1029/2009JA014392

67. Biberman L M, Vorobiev V S, Iakubov I T. Kinetics of Nonequilibrium Low-Temperature Plasma. Moscow: Nauka; 1982

68. Golubkov G V, Ivanov G K. Rydberg States of Atoms and Molecules and Elementary Processes with their Participation. Moscow: URSS; 2001

69. Radtsig A A, Smirnov B M. Handbook of Atomic and Molecular Physics. Moscow: Atomizdat; 1980

70. Noble C N, Burke P G. R–Matrix Calculations of Low-Energy Electron Scattering by Oxygen Molecules. Physical Review Letters 1992; 68(13) 2011-2014, DOI: 10.1103/PhysRevLett.68.2011

71. Sun W, Morrison M A, Isaacs W A, Trail W K, Alle D T, Gulley R J, Brennan M J, Buckman S J. Detailed Theoretical and Experimental Analysis of Low-Energy Electron-N2 Scattering. Physical Review A 1995; 52(2) 1229-1256, DOI: 10.1103/PhysRevA.52.1229

72. Greenhow C, Smith W V. Molecular quadrupole moments ofN2, andO2. Journal of Chemical Physics 1951; 19(10) 1298-1300, DOI: 10.1063/1.1748017

73. Zeiss G, Meath W J. Dispersion Energy Constants ofC 6 (A, B), Dipole Oscillator Strength Sums and Refractivities for Li, N, O, H 2, NH 3, H 2O, NO, N 2O. Molecular Physics 1977; 33(4) 1155-1176, DOI: 10.1080/00268977700100991

74. Soven P. Self-Consistent Linear Response Study of the Polarizability of Molecular Nitrogen. Journal of Chemical Physics 1985; 82(7) 3289-3291, DOI: 10.1063/1.448227

Chapter-2

Receiver Biases in Global Positioning Satellite Ranging

1. Introduction

The ionosphere is a dispersive medium, and taking advantage of this property enables the ionosphere's total electron content (TEC) to be estimated from the differential delay between the two GPS frequencies (L1 = 1575.42 and L2 = 1227.6 MHz). Other effects in the system, referred to as biases, introduce delays between the two frequencies that must be removed to resolve the TEC accurately. These biases are introduced either by the satellite or by the receiver. This chapter presents the results of an investigation into GPS receiver bias dependence on temperature. Receiver and satellite biases, can be significant, and if not treated correctly, can corrupt the GPS total electron content (TEC) estimate. It has been shown that the GPS satellite biases are fairly stable day-to-day (Sardon et al., 1994 and Wilson et al., 1999), whereas the GPS receiver biases can have significant variation, sometimes over intervals of hours [Coco, 1991]. The Jet Propulsion Laboratory (JPL) delivers estimates of the 10 day average of individual satellite biases to the U.S. Air Force. According to Komjathy (2012), the 10-day average for each satellite differs from the individual day values by 0.1-0.3 nanoseconds (ns). However, to successfully make this 10-day average measurement, the Faraday rotation due to the ionosphere must be removed with an extremely high degree of accuracy. Standard errors for absolute GPS TEC estimation are estimated by the community to be ~3 TECU (1 TECU = 10^{16} electrons/m^2), although the differential GPS TEC measurement is far more accurate (~0.02 TECU). These levels of TEC biases can pose some significant challenges in ionosphericimage reconstruction as reported by Cornely (2003). Most scientists working in the area think we should have absolute TEC values from GPS at the 0.1- 0.2 TECU level, an extremely demanding goal. Slant TEC estimates are obtained from GPS observations using the following equation Dyrud (2008):

$$STEC_{i,k} = \left[2.85\left(D_{L2}ns - D_{L2}ns - B_k - B_i + \varepsilon\right)\right]_{i,k} \tag{1}$$

In (1), $STEC_{i,k}$ represents the slant TEC and has the units of TECU (1 TECU = 10^{16} electrons/m^2), the constant 2.85 is the conversion factor from time in ns to TECU and has the units of TECU/ ns; D_{L1} and D_{L2} represent the total measured range delay in ns from the L1ε and L2 (L1 = 1575.42 and L2 = 1227.6 MHz) frequencies, respectively. The subscript k represents the satellite transmitting the data and the subscript i represents the receiver that is collecting the data. Bk and Bi represent differential delays in ns, B_{kL1}-B_{kL2} and B_{iL1}-$B_{iL2,}$ due to the satellite and receiver. Finally, ε represents other error sources, again in ns, which are assumed to be small.

It is worth noting that part of the GPS modernization includes a third civilian frequency, L5 = 1176.45 MHz, (Kaplan 2006), which will allow for additional estimates of the TEC using frequency combinations at each time sample. The advantages for the atmospheric science community for GPS modernization are described in Van Dierendonck and Coster (2001). In the early days of GPS, the community was uncertain how to estimate either set of biases. However, it was quickly recognized that the delay introduced by the ionosphere changed. Techniques applying this property to estimate the biases are described in works by Mannucci et al., 1993, and Komjathy, 1997. Work at the NASA JPL (e.g. Lanyi and Roth, 1988) led to initial algorithms for bias estimation. Other methods were developed using single receivers (Coco et al., 1991, Gaposchkin and Coster, 1993). More powerful techniques based on the global network of receivers (Wilson and Mannucci, 1993 and Sardon et al., 1994) then followed. The estimation of satellite and receiver biases remains an area of significant and active investigation within the ionospheric community. New and enhanced techniques have been recently developed that estimate receiver differential biases for all available GPS stations (typically around 1000 sites) on a daily basis (Komjathy et al., 2005 and Rideout and Coster, 2006). The research community needs more efficient and improved estimation algorithms to properly perform process and quality checks on the large amount of GPS data currently available on a daily basis.

This chapter looks at one possible error in the standard practices of determining receiver biases. Specifically, the most common procedure within the community involves applying a single value for the receiver bias, and another one for each individual satellite bias, over a 24-hour time frame. For the satellite biases, most groups rely on those estimated by the international GNSS organization (IGS) or those of IGS contributing members. These values are available on- line at the Crustal Dynamics Data Information System (CDDIS). Receiver biases, on the other hand, are typically estimated by the individual groups processing TEC. Typically, the estimate of the receiver bias is made over a full 24-hour period. This practice averages over any temporal changes in temperature. In the literature, there have been hints of issues with temperature, for example the thermal influence on time and frequency transfer has been studied (Rieck, C., 2003). It is assumed that the temperature dependence in the L1-L2 receiver bias is due to a combination of the following: 1) the hardware in the pre-amplifier of the antenna, 2) to the cable connecting the antenna, and 3) the receiver hardware itself. Receivers that measure the biases due to the receiver hardware have been built (Dyrud et al., 2008), but they neglect to account for the receiver bias due to 1) and 2). In this chapter, our aim is to demonstrate that changes in the estimated receiver bias exist due to changes in temperature, and this will affect the absolute TEC that can be obtained with GPS

2. Receiver bias estimation

To test our hypothesis of temperature dependence on receiver bias estimation, we estimated daily receiver bias values during the night-time period from 12:00 am to 2:00

am when the ionosphere is expected to be fairly stable. A software package known as MIT Automated Processing of GPS (MAPGPS) has been developed to automate the processing of GPS data into global total electron density (TEC) maps. Observations are used from all available GPS receivers during all geomagnetic conditions where data has been successfully collected. In this section, the architecture of the MAPGPS software is described. Particular attention is given to the algorithms used to estimate the individual receiver biases. The MAPGPS approach to solving the receiver bias problem uses three different methods: minimum scalloping, least squares, and zero-TEC. A top level block diagram illustration of the MAPGPS system is shown in **Figure 1**.

To estimate receiver biases, MAPGPS uses a procedure which relies on a combination of three different methods. The first method, minimum scalloping, was developed by P. Doherty (private communication) and depends on the assumption that the "scalloping" in the values of zenith TEC estimates from the different satellites observed over a 24 h period should be minimized. The second method, least squares, uses a least squares fitting routine to compute the differential receiver biases. The least squares method in MAPGPS is based on an earlier procedure developed at MIT Lincoln Laboratory (Gaposchkin and Coster 1993; K. Duh, private communication). The least squares method measures only the differential biases, however, and so must be combined with the other methods to determine absolute bias levels. The third method used by MAPGPS is the zero TEC method. In this method, the bias value of the receiver is selected to be that which sets the minimum value of the TEC over a 24-h period to be zero. We describe these methods in more detail in the following paragraphs.

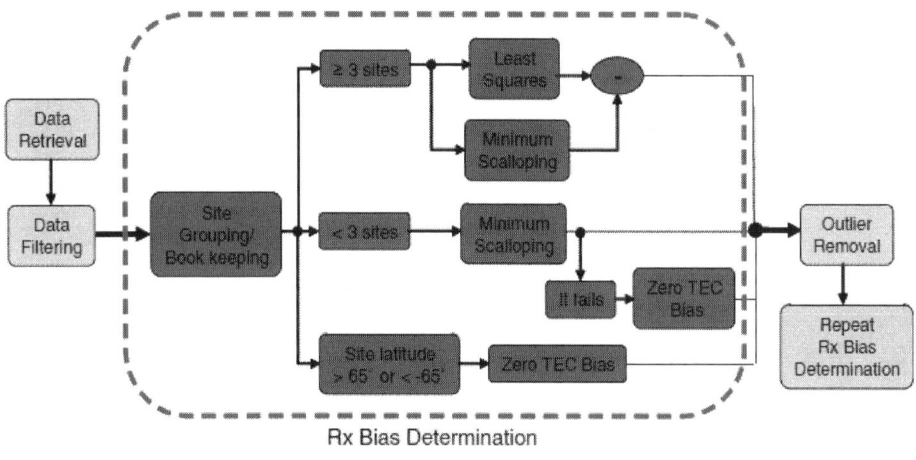

Figure 1. Illustration of the MAPGPS processing steps

3. Minimum scalloping method

The minimum scallop technique is based on the principle that zenith TEC values computed from low elevation angle data should not, on average, be different from zenith TEC values computed from high elevation angles. Of course, during some time periods, there may be temporary ionospheric structures that make this assumption false, such as at sunset or sunrise when gradients are frequently present in the TEC. Because of this, this technique has been implemented in a way that tries to average TEC values over a relatively long period. In general, a mapping function converts the line-of sight TEC at lower elevation angles to its vertical or zenith value. However, the part of the TEC observation that is caused by receiver bias, and not the ionosphere, is unaffected by elevation angle. When the mapping function is applied to a receiver using an incorrect receiver bias, the estimated vertical TEC values will be in error. Specifically, the vertical TEC values that correspond to low elevation angles will be either increased or decreased, depending on the sign of the bias. The minimum scallop technique can be applied to the TEC data around local midnight, where there should be less inhomogeneity in the TEC. For a given receiver bias, we bin and median filter the vertical TEC values by elevation angle, and determine the flatness of the resulting median TEC versus elevation-angle. The receiver bias that gives the flattest value of TEC versus elevation angle is the minimum scallop receiver bias. While no systematic errors were found to be associated with this technique, the standard deviation of day-to-day estimations of the same receiver is about 2 TECU. It is not clear whether this 2 TECU value is due to innate problems with the algorithm or with actual receiver bias variability, or some combination of both.

4. Least squares method

To use the least squares method, a system of equations is set up using a number of observations and unknowns. On any given day, each GPS receiver observes multiple satellites multiple times. Each observation consists of a TEC estimate, a satellite bias, and a receiver bias. The receiver bias is assumed to be constant over the day for each receiver, and the satellite bias is assumed to be constant over the day for each satellite. In this processing, the satellite biases are assumed to be known. The problem of estimating the receiver biases is thus over-deter- mined to a large degree as there are many more observations than unknown receiver biases. Accordingly, the method of least squares is well suited to find the best-fit solution. To compute the differential relationship between the different receiver biases, a system of difference observations is created as follows: An observation, O, is described as (ignoring measurement error):

$$Q = O - S = T + R. \tag{2}$$

where T is the line-of-sight TEC, S is the satellite bias, and R is the receiver bias. S is assumed to be known a priori from previous calculations, so it is subtracted as a constant, leaving the value Q, our partially corrected observation:

$$\mathrm{O} = \mathrm{T} + \mathrm{S} + \mathrm{R}. \tag{3}$$

If there are two observations by two different receivers, we have the equation:

$$Q_1 = Q_2 = (T_1 - T_2) + (R_1 - R_2) \tag{4}$$

Applying the vertical TEC mapping function, Z, we get:

$$\frac{Q_1}{Z_1} - \frac{Q_2}{Z_2} = \left(\frac{T_1}{Z_1} - \frac{T_2}{Z_2}\right) + \left(\frac{R_1}{Z_1} - \frac{R_2}{Z_2}\right) = \left(vT_1 - vT_2\right) + \left(\frac{R_1}{Z_1} - \frac{R_2}{Z_2}\right), \tag{5}$$

where vT represents the vertical, or zenith, TEC. If the ionosphere pierce points of the two observations are sufficiently close together, the zenith TECs cancel out and we are left with the desired equation:

$$\frac{Q_1}{Z_1} - \frac{Q_2}{Z_2} = \left(\frac{R_1}{Z_1} - \frac{R_2}{Z_2}\right) \tag{6}$$

The system equations (2)-(6) can be solved by the least squares method. MAPGPS define the observations to be sufficiently close together if their ionospheric pierce point locations at 450 km were separated by no more than 100 km in the horizontal direction. The least squares method does not determine the absolute value of the biases. It only gives the relative biases.

5. Zero TEC method

This method is based on the principle that the TEC often approaches zero during the night or at high latitudes. Therefore, the receiver bias in this method is calculated by setting the minimum observed value of the TEC over a 24 h period equal to zero. This method has the advantage that it is simple to use, and in a relative sense it is reasonably robust. Nevertheless, especially in the equatorial regions, one does not anticipate the minimum value of the TEC to be equal to zero. Noise can also affect the estimation. Using this assumption to calculate the biases may cause non-negligible errors.

6. MAPGPS receiver bias determination

In MAPGPS, the receiver bias determination is an iterative process involving several steps. This process is illustrated inside the dashed line in Figure 10. This Figure shows how the three methods described above are applied either in combination or individually to resolve the receiver bias. The first step in this process is to filter out data with elevation angles of less than 30degrees in order to separate out mapping function effects from the calcula- tion of receiver bias. This is done because scalloping can be caused by both the receiver bias and a faulty mapping function. Above 30 degrees elevation angle, all the mapping functions are very similar, and yet at 30 degrees the slant delay is still almost twice the zenith delay at zenith. This leaves enough elevation angle effect to separate out the receiver bias. Once the data with elevation angles greater than 30 degrees has been collected, groups of sites are created by starting with a randomly selected seed site. Next all remaining available receiver sites are searched to find the receiver site in the closest proximity to any of the group members. If this identified site is within a 400 km horizontal distance, it is added to the group. The 400 km horizontal distance was chosen in order to ensure that within any group there would be a large number of coincident TEC measurements (defined as TEC estimates from different receivers that are separated by not more than 50 km in the horizontal location at the pierce point height of 450 km). This procedure is repeated until either there are no remaining sites within 400 km of any group member, or the group reaches a maximum number of 100 sites. Once this step is complete, the method for determining the differential bias is dependent on the number of sites in the group. If the group consists of three or more sites, the differential biases are found using the least squares method which outputs relative, not absolute, biases. To set the level of the absolute bias, we could simply use the minimum scallop receiver bias for a single receiver. However, the minimum scallop technique has a certain amount of random uncertainty embedded in it. To reduce this uncertainty, we determine the minimum scallop receiver bias for all of the receivers in the group. We then find the average difference between these values and those computed by the least squares technique. In this way, the error in the absolute value of the receiver bias in the group is reduced by a factor of the square root of the number of receivers. Processing data from 1,000 stations takes approximately 12 h to run on a single computer with a Pentium 4 class.

If the group contains fewer than three sites, the differential bias is found using the minimum scalloping technique. First, the zero TEC bias is computed and used as the initial guess. Next for a range of biases about the initial guess, a histogram of average TEC values versus elevation angle bin is constructed for data in a 4 h window centered on 2 a.m. local (solar) time. The slope of this histogram is calculated. If the minimum slope is at the edge of the range of biases, the process is assumed to have failed and the process is repeated with the histogram con- structed for data in a 4 h window centered on solar noon. If the process fails a second time, it is repeated again with larger bias limits. In general, this process succeeds approximately 95% of the time. The zero TEC method is employed in two cases. This method is used in the 5% of the cases where the minimum scalloping technique fails and it is automatically used for receivers located at latitudes above 65 degrees latitude. The reason for this latitude criterion is the typical high latitude variability of the TEC. These values are expected to be larger in magnitude than the scalloping effect. Because of this, the minimum scalloping technique fails to work for high latitude data.

7. Description of experiments

In the experiments, four Scintillation Network Decision Aid (SCINDA) GPS receivers were used at two different sites. The SCINDA is a network of ground-based receivers that monitor scintillations at the UHF and L-Band frequencies caused by electron density irregularities in the equatorial ionosphere (Groves et al., 1997). The four SCINDA receivers are all NovAtel dual frequency GISTM receivers running the specialized software described in Carrano and Groves (2006). The temperature data were collected with EasyLog USB data loggers, which collect temperature, dew point and humidity data. Temperature data were verified by comparing the measurements to the National Oceanic and Atmospheric Administration's measurements at the site of one of the first site. These temperature loggers have a stated manufacturers accuracy of ±0.5°C(±1°F), and a repeatability of ±0.1°C (±0.2°F) over the range -35°C to 80°C (-31°F to 176°F). In the sampling, the temperature is measured to the 0.5°C level. For the experiment at the first site, data were collected for over a hundred days using two temperature sensors (one indoors and the other outdoors) and a single GPS receiver. In this experiment, the goal was to determine and remove the temperature dependence of the receiver bias. A second experiment was conducted at the other site which involved a comparison of TEC estimates from three collocated GPS receivers over a single 24-hour period. The primary goal of this experiment was to observe if the temperature dependence of receiver bias estima- tion varies from receiver to receiver.

We will first describe experiments at the first site where data was collected from day 15 to 150, 2010. This data set is used to attempt to model and remove the temperature dependence on receiver bias estimation. The remaining residual bias is examined for dependence on other parameters that may need to be considered in future modeling efforts.

8. Experiment at the first site

In a first experiment, temperature data were collected with two temperature loggers that were collocated with the single GPS SCINDA receiver that ran continuously at the first site. One of these temperature loggers was located inside the building near the receiver and the other was located outside the building on the pole of the antenna. It is important to note that the antenna was located on a black top roof, where temperatures are expected to exceed the local ambient temperature.

9. Results

For each day, an average temperature for the night-time period from 12:00 am to 2:00 am was computed. Figure2 shows the outdoor and indoor temperature data plotted as a function of the corresponding bias value from day 15 to 150 2010 (January 2010 to June 2010).

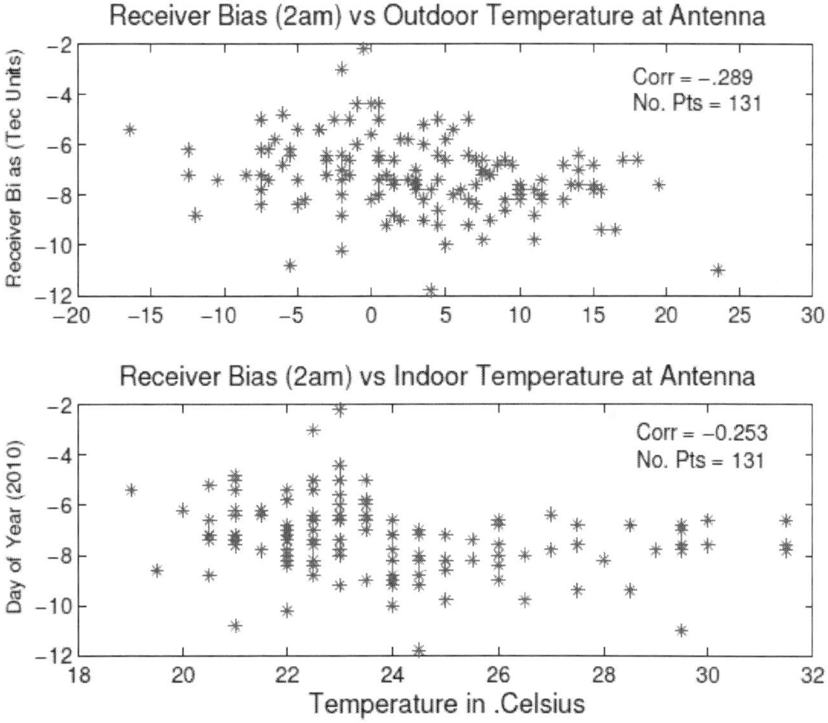

Figure 2. Night-time receiver bias versus outdoor temperature (top) or indoor temperature (bottom) for the time period day 15-150, 2010. Data collected at the MIT Haystack Observatory Optics building in Westford, MA.

It is important to note that the temperatures and bias values shown in this plot and used in this chapter are from the 12:00 am to 2:00 am local time period. The Pearson correlation coefficient (r), which represents the linear relationship between two variables, is determined for both the outdoor temperature versus bias (top plot) and the indoor temperature versus bias (bottom plot). Other fits to the data (such as fitting to a quadratic) were no more significant, as such the linear model was used exclusively. For a statistical sample size greater than 100, as is the case in the data being used with 131 points, an r value greater that.254 or less than -. 254 is rated significant. Because the Pearson correlation coefficient r, is equal to -.289 for the outdoor temperature versus bias value versus -.253 for the indoor temperature versus bias, we decided to first remove the correlation observed between the receiver bias and the outdoor temperature. In addition, as shown in **Figure 2**, there is an apparent break in the indoor temperature data versus receiver bias (this can be observed visually at approximately 24°C).

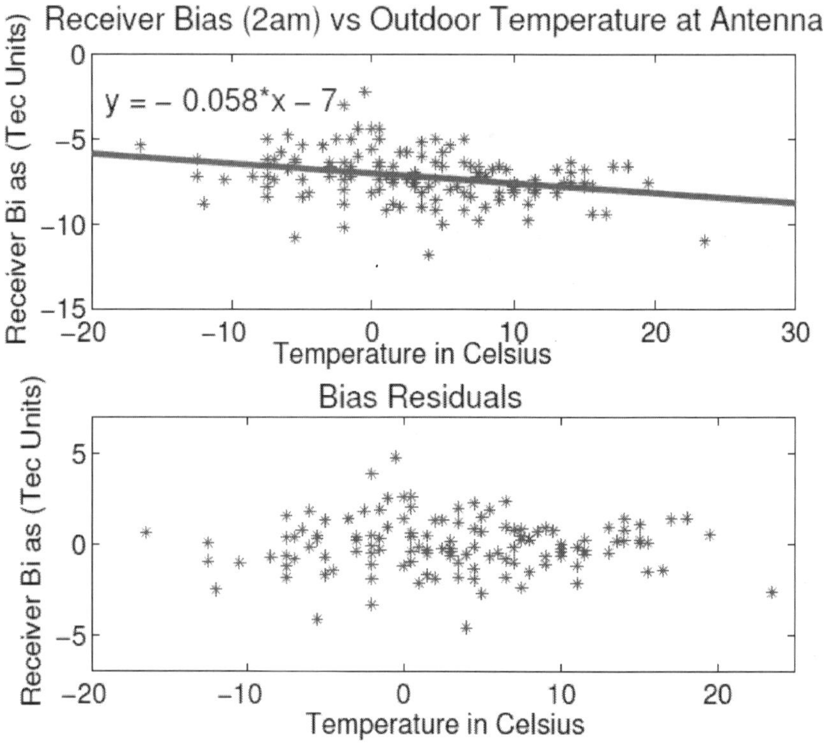

Figure 3. The top plot shows the computed linear trend in the receiver bias versus outdoor temperature data set. The bottom plot shows the receiver bias residuals versus temperature.

To remove this correlation, a linear trend was computed for the outdoor temperature versus receiver bias (shown in the top plot of Figure 2) and removed from the data. The residuals are shown in the bottom plot of Figure 2. Figure 3 shows the resulting receiver bias residuals versus the indoor temperature. In this plot, a clear break can still be observed in the data at 24°C. It is assumed that data below 24°C is indicative of time periods where the heater in the optics building was having difficulty maintaining a constant temperature. This may account for the larger fluctuations below 24°C. For this data set, the data were divided into two groups corresponding to the residuals associated with the data less than and greater than or equal to 24 °C.

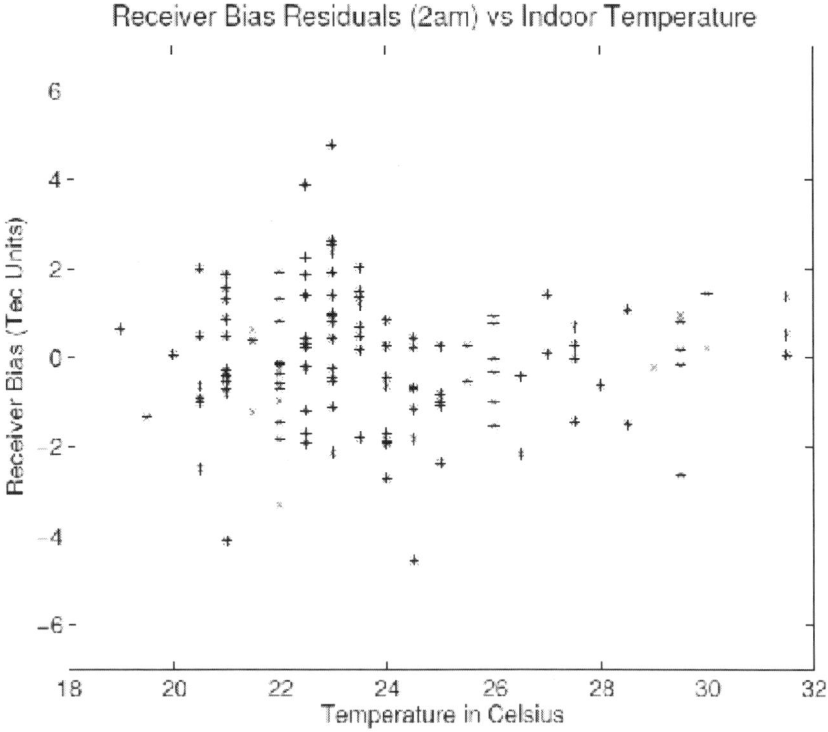

Figure 4. Receiver Bias Residuals versus Indoor Temperature for day 15-150, 2010.

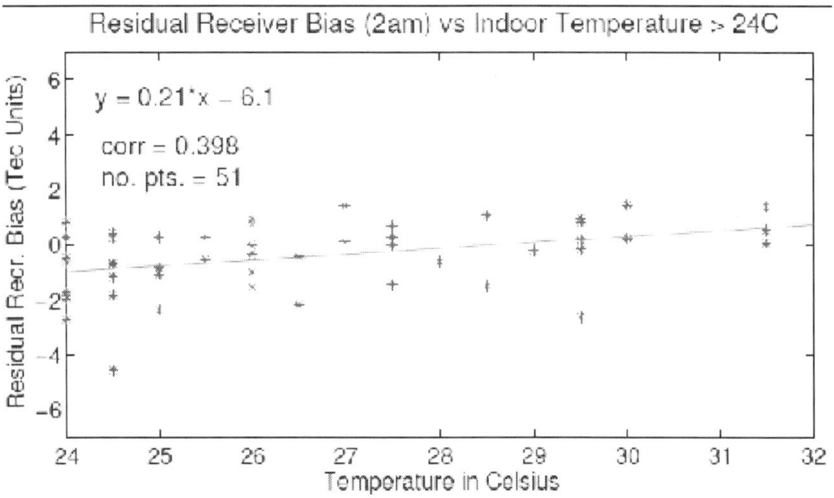

Figure 5. The computed linear trend in the receiver bias residual versus indoor temperature greater than or equal to 24°C.

Linear fits were computed for each of these groups. These fits and the corresponding Pearson r values are shown in **Figures 4** and **5**. For the data corresponding to temperatures greater than 24 °C, the Pearson r value is fairly high (r = 0.398). Once these linear trends have been estimated and removed from the data, the final residuals are shown in

Figure 6. Clearly a 2-3 TEC fluctuation level remains in the biases versus day of year plot. To test if additional factors should be accounted for in our bias estimation techniques, several atmospheric parameters were examined to see if any additional correlations could be found. The atmospheric parameters selected to study for correlations included the Ap, Kp, and the daily and 90-day averaged F10.7 cm solar flux. These parameters are important in measuring solar activity (daily and 90-day averaged F10.7 cm solar flux) and the Earth's magnetic and electric field disturbances (the Kp and Ap).

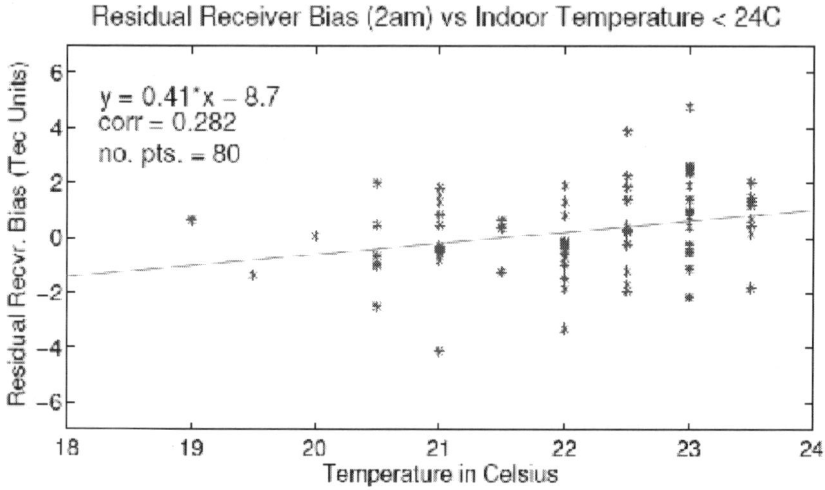

Figure 6. The computed linear trend in the receiver bias residual versus indoor temperature less than 24°C.

The only parameter that strongly stood out as highly correlated with our residuals was the daily f10.7 cm solar flux, which is a measurement of solar radio emissions at 2800 MHz. This correlation can be observed in **Figure 7** where a scaled daily F10.7 cm value is over

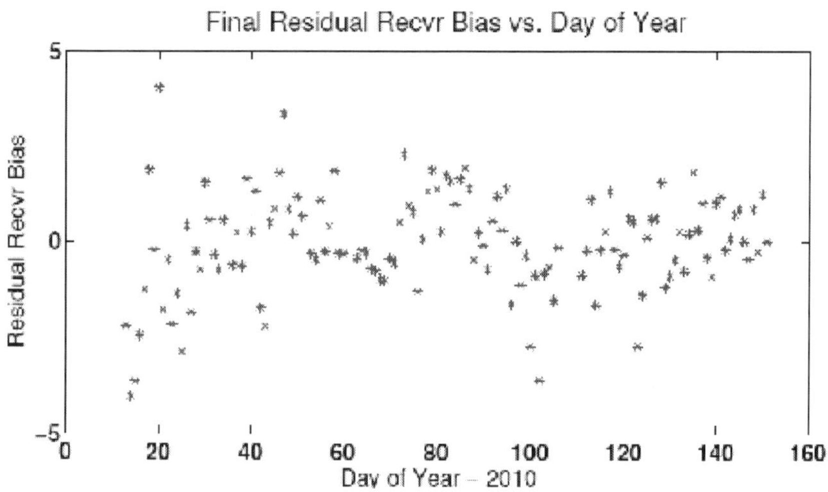

Figure 7. Final Residual Receiver Bias following removal of both indoor and outdoor temperature trends

plotted onto the final residual receiver bias. The correlation factor between these two data sets is 0.64. The observed correlation is suspected to be related to the mapping function used in the receiver bias estimation. This mapping function is described in Rideout and Coster, 2006 and depends only on the elevation, and has no built-in term for the ionospheric shell-height. This approach was based on the assumption that there is little variability in the ionospheric shell-height in the two hour window after midnight when all the receiver biases are estimated. Nevertheless, it is important to point out that the ionospheric shell-height does have some day-to-day variability that is not captured in the model being used, and this additional error source could perhaps account for some of the un-modeled errors observed. Some of this error can also be attributed to remaining errors in the satellite biases.

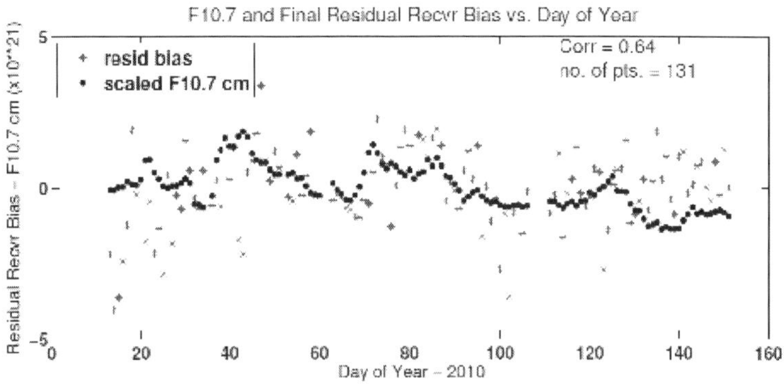

Figure 8. Receiver bias residuals and daily f10.7 measurements versus day of year 2010 as a function of time of day.

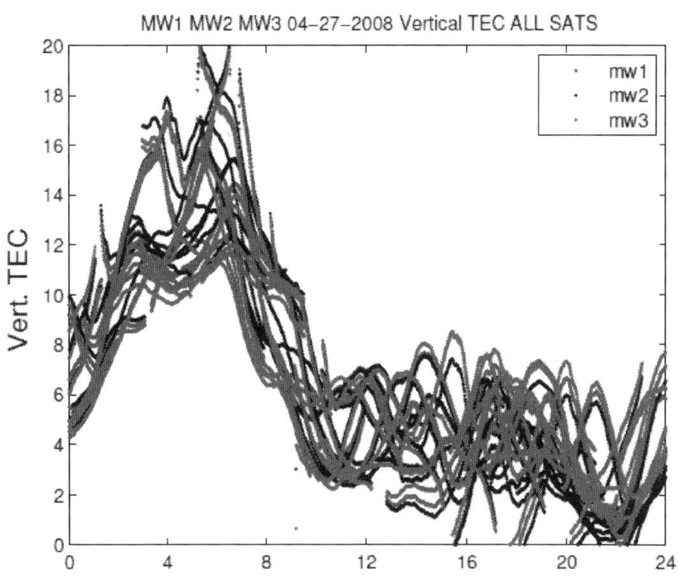

Figure 9. Vertical TEC estimates of three closely located (~10 - 20m separation) SCINDA receivers at the MWA site plotted as a function of time of day for April 27, 2008.

10. Experiment at the second site

On April 25, 2008, three Scintillation Network Decision Aid (SCINDA) GPS receivers were located at the MWA site, separated by approximately 10-20 m. *Figure 8* shows the total vertical TEC estimated by each receiver over the full 24 hour period. The vertical TEC plot is shown here to illustrate the diurnal pattern of the TEC., and indeed a clear diurnal pattern can be observed, with the maximum TEC reaching 18-20 TEC units at around 4-5 UT and the minimum TEC reaching 0-2 TEC units around 16-17 UT. Receiver bias and satellite biases have been applied to the data. The different colors (red, black, and blue) represent the TEC estimates of the different receivers. Only where the estimates of vertical TEC differ is it possible to see the different colors.

Shown in **Figure 9** are the line-of-sight (los) differential TEC measurements between these different receivers as a function of time of day. The top plot shows the differential los TEC measurements between receivers MW2 and MW1, the middle plot shows the differential los TEC between receivers MW1 and MW3, and the bottom plot show the differential los TEC between MW2 and MW3. The different colors in each plot represent the different GPS satellites. As all of receivers were within 10-20 m of each other, the differential los TEC values should be either constant as a function of time of day, or at the very least, relatively constant. What is observed, however, are larger values in the differential TEC during the daytime and smaller values during the night. This diurnal structure is largest in the bottom plot which shows the differential los TEC values between MW2 and MW3 and smallest in the middle plot which shows differential los TEC between MW1 and MW3. One would surmise from this that receiver MW2 has a receiver bias temperature dependence that differs significantly from that of MW1 and MW3. While the exact temperature excursion between day and night on April 27, 2008 is not known for the site, it is known from historical records that the low local temperature for April 27, 2008 was 68F, and the high temperature was 90F. Based on the observations in Figure

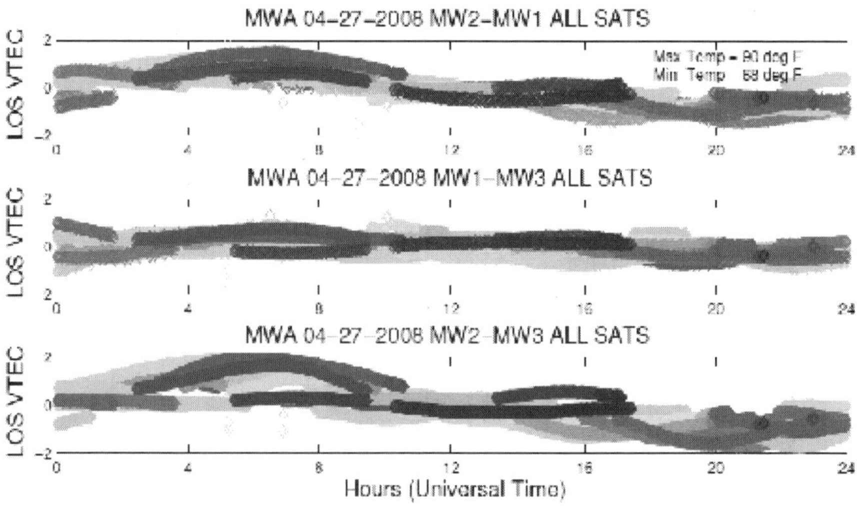

Figure 10. Differential line of sight TEC between three different SCINDA receivers at the MWA site plotted

9, we are suggesting that the observed changes in the differential TEC are due to a receiver bias temperature dependence that differs between the receivers MW1, MW2, and MW3. In other words, the temperature dependence for each receiver is unique and thus requires that the receiver bias be estimated for each individual receiver.

11. Summary

There are four main conclusions from the above discussion.

1. A clear temperature dependence on the GPS receiver bias is evident
2. This temperature dependence appears to include both indoor and outdoor temperatures
3. Other factors that are not now accounted for, such as the daily F10.7 cm flux, appear to be important in receiver bias estimation.
4. GPS receiver biases appear to be dependent on the individual receiver, and each receiver must be treated independently

With respect to how accurately the absolute total electron content can be determined using GPS for use in the low frequency array calibration it is possible that a 2-3 TEC scatter still remains in the bias estimation process even after the temperature effect is removed (see **Figure** 7). This number does not represent the relative TEC accuracy which is based on the differential phase measurements of TEC. The relative TEC can be estimated on the order of 0.01 TEC units. The 2-3 TEC scatter remaining in the observed bias estimation is likely due to the inaccuracies of the model used for the ionospheric mapping function. As was mentioned earlier, the mapping function being used does not account for any changes in the ionospheric scale height.

The error in the mapping function also lacks a valid plasmaspheric model. Recently Carrano et al., 2009 developed a Kalman filter model to estimate plasmaspheric total electron content. If a better understanding of how to model the temperature effect of the bias estimation can be obtained, it is possible that an extended version of the Kalman filter for bias determination will lead to more accurate estimates of the bias. Clearly more experimental and modeling studies are warranted.

References

1. Charles, S. Carrano and Keith M. Groves, ((2006). The GPS Segment of the AFRL-SCINDA Global Network and the Challenges of Real-Time TEC Estimation in theE-quatorial Ionosphere, Proceedings of ION NTM January 2006, Monterey, CA 1036., 2006, 18-20.
2. Carrano, C. S, Anghel, A, Quinn, R. A, & Groves, K. M. Kalman filter estimation of plasmaspheric total electron content using GPS, ((2009). Radio Science, doi: 10.1029/2008RS004070., 44RS0A10, 2009.
3. Coco, D. S, Cokcr, C. E, Dahlkc, S. R, & Clynch, J. R. (1991). Variability of GPS Satellite Differential Group Delay Biases, IEEE *Transactions* of Aerospace and Electronic Systems, 27 (6), 931- 938.
4. Cornely, P-R.J. Flexible Prior Model: Three-dimensional ionospheric tomography, Radio Sci., doi:10.1020/2002RS002703,5

5. Coster, A, & Komjathy, A. (2008). Space Weather and the Global Positioning System, *Space Weather*, 6, S06D04, doi:10.1029/SW000400.

6. Dyrud, L, Jovancevic, A, Brown, A, Wilson, D, & Ganguly, S. (2008). Ionospheric Measurement with GPS: Receiver techniques and methods, *Radio Sci.*, 43, RS6002, doi:10.1029/2007RS003770.

7. Gaposchkin, E. M, & Coster, A. J. Bias Determination, Lincoln Laboratory Technical Report 971 (MIT)., 1-1.

8. Groves, K, Basu, S, Weber, E, Smitham, M, Kuenzler, H, Valladares, C, & Sheehan, R. MacKenzie, E., Secan, J., Ning, P., McNeil, W., Moonan, D., and Kendra, M., ((1997). Equatorial scintillation and systems support, Radio Sci., 32, 2047.

9. Kaplan, E. D. Understanding GPS: principles and applications, ((2006). nav- techgps. com, Artech House, Norwood, MA.

10. Komjathy, A. (1997). Global Ionospheric Total Electron Content Mapping Using the. Global Positioning System. Ph.D. dissertation, Department of Geodesy and Geomatics Engineering, University of New Brunswick.

11. Komjathy, A, Sparks, L, Wilson, B. D, & Mannucci, A. J. (2005). Automated Daily Processing of more than 1000 ground-based GPS Receivers to atudy intense ionospheric storms, Radio Science, RS6006, doi:10.1029/2005RS003279,Komjathy, A., pri- vate communication, 2012., 40

12. Lanyi, G, & Roth, T. (1988). A comparison of mapped and measured total ionospheric electron content using global positioning system and beacon satellite observations, *Radio Sci,*, 23 (4), 483-4.

13. Rieck, C, Jarlemark, P, Jaldehag, K, & Johansson, J. (2003). Thermal influence on the receiver chain of GPS carrier phase equipment for time and frequency *transfer, Frequency Control Symposium and PDA Exhibition Jointly with the 17th European Frequency and Time Forum, 2003. Proceedings of the 2003 IEEE International,* vol., no., May 2003, doi:FREQ.2003.1275110., 326-331.

14. Rideout, W, & Coster, A. (2006). Automated GPS processing for global total electron content data, GPS Solutions, DOIs10291-006-0029-5.

15. Sardon, E, Rius, A, & Zarraoa, N. (1994). Estimation of the transmitter and receiver differential biases and the ionospheric total electron content from Global Positioning System observations," Radio Science May-June 1994, p. 577- 586., 29(3), 577-586.

16. Van Dierendonck, A. J, & Coster, A. J. (2001). Benefits of GPS Modernization to the Atmospheric Science Community, Proceedings of ION 57th Annual Meeting, June, 2001, Albuquerque, New Mexico, 382-390., 11-13.

17. Wilson, B. D, Mannucci, A. J, & Edwards, C. D. (1993). Sub-daily northern hemi- sphere ionospheric maps using the IGS GPS network, *Proceedings of the Seventh International* Wilson, B., D; Yinger, Colleen H; Feess, William A; Shank, Chris, (1999), New and improved- The broadcast inter-frequency biases, *GPS World,* Sept. 1999,, 10(9), 56-66.

Chapter-3

Stand-Alone Satellite-Based Global Positioning

1. Introduction

The main principle of stand-alone positioning with pseudo-range is a "multiple intersection" based on measurement of distances [1].

If the satellite's position is known and the "range" between each satellite and the rover receiver is measured, the rover position in the same reference system of the satellite's position could be estimated.

In this chapter, principles of measurement and least squares calculations will be discussed, with the purpose of estimating the receiver position and its accuracy. This accuracy can be "foreseen" through the DOP calculation [2], [3].

The aim of this chapter is not to show new procedures but to explain concepts described in the bibliography, often fragmentary and without examples. To improve comprehension of the values and concepts, some numeric examples are given.

2. Satellite positions

Each application of GPS positioning (stand-alone, relative, differential, generation of differ- ential correction, etc) is based on knowledge of a satellite's position in ECEF (Earth Centred Earth Fixed) geocentric cartesian reference system, which is fixed to the Earth.

The satellite position is estimated using the ephemerides which are contained in the navigation message broadcast by the satellite.

In the case of the GPS system, the ECEF satellite position is not included in this message, where dedicated parameters are contained which, following Keplerian law and considering the perturbation effect, allow calculation of the satellite position [6].

GPS ephemerides allow for the estimation of the satellite position in an inertial reference system, from which it is possible to define the ECEF coordinates.

This system is described in Fig. 1 where:

V is the vernal point, the intersection of the ecliptic and the equatorial planes.

X^0 is the axis connecting the Earth's origin C and the point V. This direction is considered fixed, excluding some variations due to secular effects.

In the ECEF system, X is the axis of intersection of equatorial plane and the Greenwich mean meridian plane. Z is the axis which coincides with the Earth's rotation axis and Y completes a right-handed orthogonal system.

The Greenwich meridian plane rotates around the Earth's rotation axis with an angular speed ω_e =7.2921151467*10^{-5} rad/s, in accordance with GAST (Greenwich Apparent Sideral Time).

Θ, the subtended angle between X and X^0, is "sideral time".

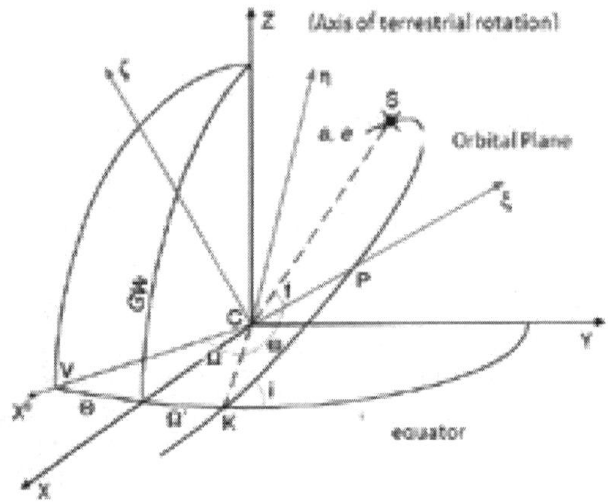

Figure 1. ECEF system and Keplerian elements

The six parameters used to describing the satellite orbit are given in Table 1.

Table 1. Keplerian elements

a	Semi-major axis of the elliptical orbit	Size and shape of the orbit Orbit
e =	orbit eccentricity (b= semi minor axis)	
ω	Argument of periapsis	Orbital plane in the apparent system
Ω	Longitude of ascending node	
i	Orbit inclination	
μ_0	Mean anomaly at the reference epoch t_{oe}[1]	Position in the orbital plane

[1]*The Keplerian element is t_0, which is the epoch of the passage to periapsis. In the GPS navigation message, the anomaly at the ephemerids reference epoch t_{oe} is given.*

where:

- Ω is the longitude of the ascending node, it is the angle which has its vertex in the centre of the Earth and it is subtended by X^0 and the ascending node K.
- P is the periapsis and it is the closest position of the satellite with respect to the Earth. The angle in C between the ascending node K and P is the periapsis argument.
- The angle i between the equatorial plane and orbital plane is the orbit inclination.
- f is the "true anomaly", shown in Fig. 1.

a and e^2 are known, the satellite position can be estimated starting from an initial position in the orbital plane, which is defined by a mean anomaly μ_0. These parameters are in the RINEX navigation message (.nav), as described in Fig. 2.

```
.. omissis ….
                                                                  END OF HEADER
23 12  4  6  7 59 44.0  .217401422560D-03 -.352429196937D-11  .000000000000D+00
     .400000000000D+01 -.695937500000D+02  .486913139030D-08 -.268627840472D+01
    -.362098217010D-05  .768281961791D-02  .245310366154D-05  .515364717865D+04
     .460784000000D+06 -.124797224998D-06  .187886879701D+01 -.838190317154D-07
     .960393629484D+00  .333250000000D+03 -.302654843351D+01 -.847749597888D-08
    -.289297764699D-09  .100000000000D+01  .168200000000D+04  .000000000000D+00
     .200000000000D+01  .000000000000D+00 -.200234353542D-07  .400000000000D+01
     .460799000000D+06
… omissis ….
```

Figure 2. Broadcast GPS ephemerides in RINEX navigation (.nav) for satellite ID=23. Keplerian elements are reported in grey and the clock corrections are reported in the boxes.

Broadcast GALILEO ephemerides are defined in the same way [10].

GLONASS ephemerides are instead described in a different way: navigational messages in the reference system, PZ90, already contain the ECEF position, velocity and acceleration of each satellite, estimated every 30 minutes [5].

Each set of parameters can be applied in estimating the satellite's position 15 minutes before and after the time reference indicated in the navigation file.

A part of GLONASS navigation file is shown in Figure 3.

Satellite position is estimated using the 4th order Runge-Kutta numerical integration of [7].

```
… omissis ….
                                                                  END OF HEADER
 4 12  4  6  7 45  0.0  .816304236650D-05  .909494701773D-12  .287700000000D+05
     .519351074219D+04 -.131477260590D+01  .279396772385D-08  .000000000000D+00
     .123013076172D+05  .266929721832D+01  .372529029846D-08  .600000000000D+01
     .217447719727D+05 -.119645977020D+01 -.186264514923D-08  .000000000000D+00
… omissis ….
```

Figure 3. Broadcast GLONASS ephemerides are contained in the RINEX navigation file for Satellite ID 4. Geocentric positions (XYZ) are highlighted in grey, the velocity in bold and the accelerations underlined. The clock parameters are indicated by boxes.

Broadcast ephemerides have about 1 meter accuracy, but more precision may be necessary for geodetic application. Precise ephemerides can be used as an alternative to broadcast ephem- erides. These ephemerides are calculated "a posteriori", and they have a centimetric level accuracy. The SP3 format [12] is used to describe these orbits and positions and satellite clock offsets are contained both for GPS and GLONASS constellations, with a rate of 15 minutes.

An example is reported in Fig. 4 [8].

```
...  omissis ....
/* RAPID ORBIT COMBINATION FROM WEIGHTED AVERAGE OF:
/* cod emr esa gfz jpl ngs sio usn
/* REFERENCED TO IGS TIME (IGST) AND TO WEIGHTED MEAN POLE:
/* PCV:IGS08_1685 OL/AL:FES2004  NONE     Y  ORB:CMB CLK:CMB
* 2012  5 17  0  0  0.00000000
PG01  13902.865823 -22573.038946   1296.775648    256.852286  8  7 10 106
PG02 -17957.886717   3939.680710 -18860.563441    390.462830  8  7  7 123
PG03  18912.061622   2346.528316  18141.509006     49.816678  8  7  2 108
PG04 -13417.970446  -8418.306518 -21560.579499    185.598885  7  9  4 124
PG05 -26445.228860   1860.936143   2480.969865   -315.972114  8  7  7 132
PG06  19522.934772   7206.064156  16797.290205     25.640080  7 11  6 113
...  omissis ....
```

Figure 4. Ephemerids in SP3 format: PG #sat., X, Y, Z ECEF position

3. Range measurement using pseudo-range

The measurement of the range between the satellite and the receiver is estimated using the pseudo-range, the code part of the signal, which is composed of squared waves. These waves are generated by an algorithm PRN (pseudo-random noise), which is periodically repeated over time [4].

In the GPS system, this component is composed of the pseudo-range C/A and P, when available. From the IIRM block, the new pseudo-range L2C is also available.

In the GLONASS system, there are two pseudo-ranges: ST (standard accuracy) and VT (usage agreed with the Russian Federation Defence Ministry) [5].

The theoretical part of the measurement of distance between the satellite and the receiver (the range) is based on the measure of "time of fly". It is the time taken between the transmission of the signal by the satellite and the receipt by the receiver.

The range measurement is realized by means of a cross-correlation procedure between two signals. When the signal is received by the receiver, the receiver generates an identical replica and moves it with respect to time. This operation is concluded when the maximum correlation is reached (Fig. 5).

In others words, the time of fly Δt is the movement that has to be applied to the replica of the signal in order to have correct alignment with the received signal.

The two signals, received and replica, are identical, but there is a misalignment caused by the travel time in space between the satellite and the receiver, as defined in equation 1.

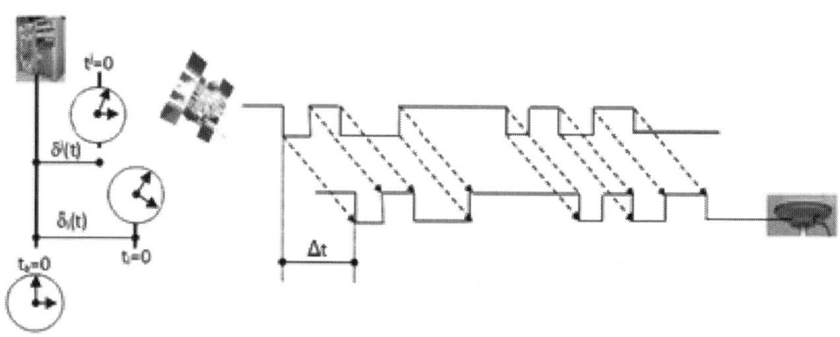

Figure 5. Pseudo-range measurement and time scale

$$R_i^j(t) = c\Delta t \tag{1}$$

where:

R_j^i is the measured range between satellite (j) and receiver (i);

c ≅ 3*10^8 m/s = light speed in vacuum.

In this way, the measured distance is a "pseudo-range" because there is an offset between the satellite clock and the receiver clock.

We consider three different time reference scales (see Fig. 5):

- atomic time scale (t_a), which is the GPS time fundamental reference maintained by the clocks in the control centre;
- satellite time scale (t_j), which is defined by the atomic clocks housed in each satellite;
- receiver time scale (t_i), which is determined by the internal receiver clock (usually a quartz oscillator).

The satellite and receiver time scales are aligned with the fundamental scale (t_a), when the offset δ^j and δ_i are estimated. These offsets are time dependent and they have to be considered as biases in the measurement of the range.

If each time is referenced to the fundamental scale, the measured range will be:

$$R_i^j(t) = c\left[\left(t^j - \delta^j(t)\right) - \left(t_i - \delta_i(t)\right)\right] = c\left[\left(t^j - t_i\right) - \left(\delta^j(t) - \delta_i(t)\right)\right] = \\ = \rho_i^j - c\left(\delta^j(t) - \delta_i(t)\right) \tag{2}$$

where ρ is the geometric range between the two phase centres of the satellite and receiver antennas, described as follows.

$$\rho_i^j(t) = \sqrt{\left(X^j - X_i\right)^2 + \left(Y^j - Y_i\right)^2 + \left(Z^j - Z_i\right)^2} \tag{3}$$

The bias due to the satellite clock is modeled with a low order polynomial function (2nd degree for GPS and GALILEO and linear for GLONASS), where the coefficients are broadcast in the navigation message.

Fig. 2 and Fig. 3 give examples of the RINEX navigation files for the GPS and GLONASS constellations.

In the GPS situation, the group delay (TGD) and the relativistic effect Δtr have to be considered, in order to estimate the satellite clock offset, using the following equation:

$$\delta^j(t) = a_0 + a_1\left(t - t_{oc}\right) + a_2\left(t - t_{oc}\right)^2 + \Delta t_r - T_{GD} \tag{4}$$

The velocity of a group is the velocity of the signal propagation and it is different from the single phase velocity of each component. TGD is broadcast in the navigation message.

The relativistic effect is due to the satellite high-speed in its orbit which has to be considered in proximity to the Earth, for its mass and its potential.

The element of relativistic correction is calculated according to GPS specification [6]:

$$\Delta t_r = F\sqrt{a}e \sin E_K \tag{5}$$

where a, e, E_K are semi-major axis, orbit eccentricity and satellite anomaly, respectively.

$F = -\frac{2\sqrt{GM}}{c^2}$ and it is defined in [6].

For example, the relativistic effect for a GPS satellite, considering: $a=26500\ km$, $e=10^{-2}\ m$, $sinE_K=1$, $F= - 4.442807633*10^{-10}\ s / m^2$ is:

$\Delta t_r = - 4.442807633 * 10^{-10} \sqrt{26500000} * 1 * 10^{-2} \approx 2 * 10^{-8}s \approx 20ns$

The synchronism between all atomic satellite clocks is considered.

It is not possible to remove the effect of satellite clock bias in stand-alone positioning as it leads to an error of about 1 ns (10^{-9} s), which corresponds to an error equal to 30 cm of the range measured between the satellite and the receiver.

The receiver clocks are typically quartz oscillators, which have less long-term stability com- pared to atomic clocks. On the supposition that errors of synchronization is approximately equal to 1 ms (10^{-3} s). This error, in the distance between the satellite and the receiver considering speed of light, is equal to 300 km.

The receiver clock offset is difficult to model therefore an additional unknown is considered at each epoch of measurement: the bias of the receiver clock δ_i.

Separating unknown and known terms in Equation 2, the pseudo-range equation will be:

$$R_i^j(t) + c\delta^j(t) = \rho_i^j(t) - c\delta_i(t) \tag{6}$$

where the unknowns are:

(X, Y, Z) position of the receiver;

δ_i receiver clock offset.

The system could be solved only if a sufficient or redundant number of equations of obser- vation is available.

4. Observation equations

We consider two different types of positioning: static and kinematic.

In the first, the receiver is stationary over a point for several epochs, without changing its position. In the second, the receiver moves and its coordinates change in each epoch. In the kinematic, four unknowns have to be solved for each epoch: three for the position (X, Y, Z) and one for the receiver clock offset [9].

Linearization of Equation (6) and considering the approximate values of the receiver position, as:

$$X_i = X_i^{(0)} + x_i;\quad Y_i = Y_i^{(0)} + y_i;\quad Z_i = Z_i^{(0)} + z_i; \tag{7}$$

the effective position unknowns will be the corrections x_i, y_i, z_i. The solution procedure is iterative and the final solution is independent of the approximate values.

If the approximate values of the position are not available, it is possible to set them equal to zero, solving the system with additional iterations.

In order to define the system, it is important to linearize the equation of the geometric range, by means a Taylor series, as follows:

$$\rho_i^j = \rho_i^{j(0)} + \left(\frac{\partial \rho_i^j}{\partial X_i}\right)_{(0)} x_i + \left(\frac{\partial \rho_i^j}{\partial Y_i}\right)_{(0)} y_i + \left(\frac{\partial \rho_i^j}{\partial Z_i}\right)_{(0)} z_i + 2nd\,order \tag{8}$$

The second order terms are small and can be neglected, and, substitute for the first derivatives, Equation (6) becomes:

$$R_i^j(t) + c\delta^j(t) = \sqrt{\left(X^j - X_i^{(0)}\right)^2 + \left(Y^j - Y_i^{(0)}\right)^2 + \left(Z^j - Z_i^{(0)}\right)^2} + \\ -\left(\frac{X^j - X_i^{(0)}}{\rho_i^{j(0)}}\right)x_i - \left(\frac{Y^j - Y_i^{(0)}}{\rho_i^{j(0)}}\right)y_i - \left(\frac{Z^j - Z_i^{(0)}}{\rho_i^{j(0)}}\right)z_i - c\delta_i(t) \tag{9}$$

which reduced to the Gauss model:

$$Ax - l_o = v \tag{10}$$

using the least squares estimator to define the solution, if the number of satellites is greater than four.

In the case of five satellites, (apexes in Equation (11)), the design matrix (A), known terms vector (l0) and unknowns (x) can be written as:

$$\begin{bmatrix} \frac{X_i^{(0)} - X^{(1)}}{\rho_i^{1(0)}} & \frac{Y_i^{(0)} - Y^{(1)}}{\rho_i^{1(0)}} & \frac{Z_i^{(0)} - Z^{(1)}}{\rho_i^{1(0)}} & -1 \\ \frac{X_i^{(0)} - X^{(2)}}{\rho_i^{2(0)}} & \frac{Y_i^{(0)} - Y^{(2)}}{\rho_i^{2(0)}} & \frac{Z_i^{(0)} - Z^{(2)}}{\rho_i^{2(0)}} & -1 \\ \frac{X_i^{(0)} - X^{(3)}}{\rho_i^{3(0)}} & \frac{Y_i^{(0)} - Y^{(3)}}{\rho_i^{3(0)}} & \frac{Z_i^{(0)} - Z^{(3)}}{\rho_i^{3(0)}} & -1 \\ \frac{X_i^{(0)} - X^{(4)}}{\rho_i^{4(0)}} & \frac{Y_i^{(0)} - Y^{(4)}}{\rho_i^{4(0)}} & \frac{Z_i^{(0)} - Z^{(4)}}{\rho_i^{4(0)}} & -1 \\ \frac{X_i^{(0)} - X^{(5)}}{\rho_i^{5(0)}} & \frac{Y_i^{(0)} - Y^{(5)}}{\rho_i^{5(0)}} & \frac{Z_i^{(0)} - Z^{(5)}}{\rho_i^{5(0)}} & -1 \end{bmatrix} \begin{pmatrix} x_i \\ y_i \\ z_i \\ c\delta_i \end{pmatrix} - \begin{pmatrix} R_i^{(1)} + c\delta^{(1)} - \rho_i^{1(0)} \\ R_i^{(2)} + c\delta^{(2)} - \rho_i^{2(0)} \\ R_i^{(3)} + c\delta^{(3)} - \rho_i^{3(0)} \\ R_i^{(4)} + c\delta^{(4)} - \rho_i^{4(0)} \\ R_i^{(5)} + c\delta^{(5)} - \rho_i^{5(0)} \end{pmatrix} = \begin{pmatrix} v^{(1)} \\ v^{(2)} \\ v^{(3)} \\ v^{(4)} \\ v^{(5)} \end{pmatrix} \tag{11}$$

It is convenient to define the offset of the receiver clock in metres, in order to avoid possible problems of ill conditioning of the system, due to the light speed c which is prevalent with respect to the other values. This is easily achievable, multiplying the offset by the light speed.

This positioning model with pseudo-range measurement defines a single position at each epoch and is useful for kinematic procedures.

The static solution requires adding additional columns in the design matrix (A), in order to estimate the offset of the receiver clock at each epoch.

To consider 2 epochs of measurement in static session, Equation (11) has to be modified as follows:

$$\begin{array}{c} \text{epoch 1} \\ (t_1) \\ \\ \text{epoch 2} \\ (t_2) \end{array} \begin{bmatrix} \frac{X_i^{(0)}-X^{(1)}(t_1)}{\rho_i^{1(0)}(t_1)} & \frac{Y_i^{(0)}-Y^{(1)}(t_1)}{\rho_i^{1(0)}(t_1)} & \frac{Z_i^{(0)}-Z^{(1)}(t_1)}{\rho_i^{1(0)}(t_1)} & -1 & 0 \\ \cdots & \cdots & \cdots & \cdots & \cdots \\ \frac{X_i^{(0)}-X^{(5)}(t_1)}{\rho_i^{5(0)}(t_1)} & \frac{Y_i^{(0)}-Y^{(5)}(t_1)}{\rho_i^{5(0)}(t_1)} & \frac{Z_i^{(0)}-Z^{(5)}(t_1)}{\rho_i^{5(0)}(t_1)} & -1 & 0 \\ \hline \frac{X_i^{(0)}-X^{(1)}(t_2)}{\rho_i^{1(0)}(t_2)} & \frac{Y_i^{(0)}-Y^{(1)}(t_2)}{\rho_i^{1(0)}(t_2)} & \frac{Z_i^{(0)}-Z^{(1)}(t_2)}{\rho_i^{1(0)}(t_2)} & 0 & -1 \\ \cdots & \cdots & \cdots & \cdots & \cdots \\ \frac{X_i^{(0)}-X^{(5)}(t_2)}{\rho_i^{5(0)}(t_2)} & \frac{Y_i^{(0)}-Y^{(5)}(t_2)}{\rho_i^{5(0)}(t_2)} & \frac{Z_i^{(0)}-Z^{(5)}(t_2)}{\rho_i^{5(0)}(t_2)} & 0 & -1 \end{bmatrix} \begin{pmatrix} x_i \\ y_i \\ z_i \\ c\delta_i(t_1) \\ c\delta_i(t_2) \end{pmatrix} - \begin{pmatrix} R_i^{(1)}(t_1)+c\delta^{(1)}(t_1)-\rho_i^{1(0)}(t_1) \\ \cdots \\ R_i^{(5)}(t_1)+c\delta^{(5)}(t_1)-\rho_i^{5(0)}(t_1) \\ \hline R_i^{(1)}(t_2)+c\delta^{(1)}(t_2)-\rho_i^{1(0)}(t_2) \\ \cdots \\ R_i^{(5)}(t_2)+c\delta^{(5)}(t_2)-\rho_i^{5(0)}(t_2) \end{pmatrix} = (\bar{v})$$

(12)

5. Least Squares

The least squares solution starts from the Gauss model, which was described in (10), and is defined by means of calculus and inversion of the normal matrix N.

The estimated residuals v^ and the variance–covariance matrix of the estimated solution C_{xx} can be calculated, considering:

$$\hat{x} = N^{-1}T_n$$

(13)

where:

$N = A^T PA$ is called normal matrix

$T_n = A^T Pl_0$.

Some consideration will given to the weight matrix P. It is generally obtained by inverting the variance–covariance matrix of the observations C_{LL} multiplied by the variance of the unit of weight, which is defined "a priori".

$$P = \sigma_0^2 C_{LL}^{-1}$$

(14)

Different methods can be used to defining the matrix P:

- *P* equal to the identical matrix I if the ranges are considered to have the same weight;
- *P* is a diagonal matrix, with different values, as in the following example.

$$P = \begin{bmatrix} p_1 & & & & \\ & p_2 & & 0 & \\ & & p_3 & & \\ & 0 & & p_4 & \\ & & & & p_5 \end{bmatrix} \tag{15}$$

But what is a possible correct weight to use in P?

There are different strategies used to selecting the weight of the range:

- Starting from the **URA** (*User Range Accuracy*), which represents the maximum error of the range foreseen during the period of validation of the ephemerides, it is a statistical value of the accuracy for each satellite. **URA** is contained in the navigation message and it is independent of the satellite's elevation or other environmental conditions. Using the GPS specification the accuracy of each satellite can be estimated [13].
- Each weight can be considered depending on the satellite's elevation. For example, a function of the zenithal angle *z* can be used, where each satellite can be weighted with the following model: p =cos2z. EUREF suggests an alternative method [11]:

  $$p = cos^2 z + a\, sin^2 z \tag{16}$$

  with a recommended value of a =0.3.

  These weight functions are shown in Fig. 6.
- Signal-noise ratio (SNR) observed each epoch for each range could be used as the weight.

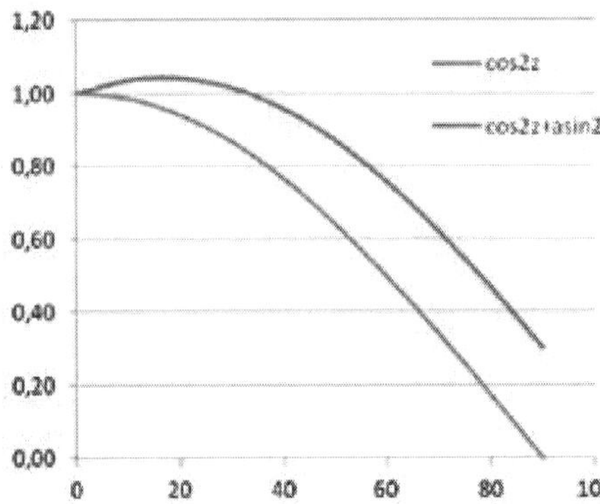

Figure 6. Weights dependence on zenithal angle

Based on the estimated solution $\hat{x}$, the estimated residuals $\hat{v}$, and the apostriori variance of the unit weight, can be obtained:

$$\hat{v} = A\hat{x} - l_0 \quad , \quad \hat{\sigma}_0^2 = \frac{\hat{v}^T P \hat{v}}{n - r} \tag{17}$$

where:

n= number of equations;

r= number of unknowns.

Finally, the variance–covariance matrix of the estimated solution *Cxx* can be determined, considering the normal matrix and the apostriori variance of unit weight:

$$C_{xx} = \hat{\sigma}_0^2 N^{-1} = \begin{bmatrix} \sigma_X^2 & & & \\ \sigma_{XY} & \sigma_Y^2 & & \\ \sigma_{XZ} & \sigma_{YZ} & \sigma_Z^2 & \\ \sigma_{Xt} & \sigma_{Yt} & \sigma_{Zt} & \sigma_t^2 \end{bmatrix} \tag{18}$$

This matrix has the variances of the estimated parameters along the diagonal, and the covariances are in the off-diagonal elements.

The matrix N^{-1} is also called the "cofactor matrix"; it is independent of the measurements, but only dependent on the geometrical satellite-receiver configuration. We will show later how N^{-1} is used to calculate DOP.

6. Measurement bias of the pseudo-range

In order to consider the several range measurement biases, Equation (6) has to be modified as follows:

$$R_i^j(t) + c\delta^j(t) = \rho_i^j(t) - c\delta_i(t) + T_i^j(t) + I_i^j(t) + E_i^j(t) \tag{19}$$

where:

T = tropospheric delay;

I = ionospheric delay;

E = ephemerides error.

A short description will be given in the following.

Tropospheric delay is an error which occurs in only the low part of the atmosphere, up to 40 km from the Earth's surface.

This error depends on: pressure, temperature and relative humidity. It can be estimated using a model, such as the one described in [3, 4]. In the standard condition (i.e. temperature =273.16 K, pression=1013.25 mbar, e=0), this error is equal to 2.3 m in the zenith position (z=0°) and it increases when the zenithal angle increases.

This is the main reason that it is better to avoid using satellites with low elevation (zenithal angle > 75°) to realize the positioning.

Ionospheric delay, on the other hand, depends on the electronic content of the ionosphere layer of the atmosphere between 40 km and 1000 km above the Earth's surface; it changes with the sun's activities. This delay is dispersive, that is, it depends on the signal frequency. The value of this delay is variable but it is normally greater than the tropospheric delay. Dual frequencies receivers can use the dispersive nature of the ionosphere to completely remove the delay.

Ephemeride errors depend on the satellite position: broadcast ephemerides have a meter-level accuracy; precise or predicted rapid products instead have centimeter level of accuracy. These products are available on the IGS (International GNSS Service) website.

These errors are spatially correlated; therefore they have similar effects on two receivers in close proximity. These biases can be eliminated or reduced using relative positioning, but not in the stand-alone positioning.

Others errors such as phase centre variation, multipath and hardware delay are less significant in this context and are not considered.

7. Relative motion in the stand-alone positioning

The position, estimated as described above, does not consider certain important effects which happen during the time of the propagation of the signal from the satellite to the receiver (about 67 ms).

The following effects occur during this period:

- the satellite position is changed by about 250 m;
- the Earth has rotated about its spin axis by about 30 m to the East on the equator.

The estimation of the receiver position requires an additional iteration where the satellite position is modified, taking into account the propagation time τ.

$$t^j = t^j - \tau_i^j \qquad \tau_i^j = \rho_i^j / c \tag{20}$$

ρ^i_j can be estimated using the satellite coordinates and the first approximate position of the receiver. The effect of the Earth rotation is considered, applying the velocity of terrestrial rotation ωe to the coordinates *X* and *Y*, as:

$$\begin{pmatrix} X_i^j \\ Y_i^j \\ Z_i^j \end{pmatrix} = \begin{bmatrix} \cos(\omega_e \tau_i^j) & -\sin(\omega_e \tau_i^j) & 0 \\ \sin(\omega_e \tau_i^j) & \cos(\omega_e \tau_i^j) & 0 \\ 0 & 0 & 1 \end{bmatrix} \begin{pmatrix} X^j \\ Y^j \\ Z^j \end{pmatrix} \tag{21}$$

In conclusion, this position is determined by also taking into account both the time propagation and the Earth's rotation, leading to a more precise receiver position. A further iteration does not give more significant benefit in terms of accuracy.

8. DOP and satellite visibility

Sometimes it could be useful to define the ECEF coordinates with respect to a local plane, which is tangential to the ellipsoid at a defined point. This local system allows the separation of the horizontal component from the vertical, where the GPS is less precise. If this system is used, the variance–covariance matrix has also to be rotated.

Therefore, it is possible to define a local plane with its origin at P_i, with geographical coordi- nates φ and λ, considering an Eulerian Cartesian tern, defined as following:

- u-axis coincides with the normal to the ellipsoid passing through Pi;
- n-axis coincides with the meridian tangent directed north;
- e-axis completes the clockwise tern.

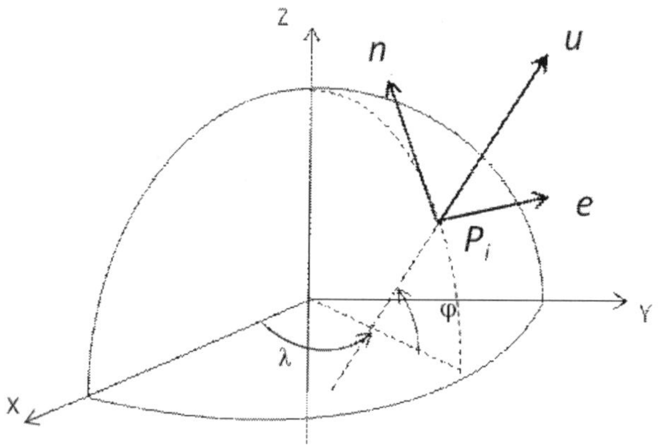

Figure 7. ECEF coordinates (X Y Z) and Eulerian system (e n u)

The P coordinates can be expressed in the ECEF system as:

$$X = (N+h)\cos\phi\cos\lambda$$

$$Y = (N+h)\cos\phi\sin\lambda \quad ; \qquad N = \frac{a}{\sqrt{1-e^2\sin^2\phi}}$$

$$Z = \left[N\left(1-e^2\right)+h\right]\sin\phi \tag{22}$$

The coordinates of the receiver and of satellite k can be roto-translated into a Eulerian tern, using the following [1, 2]:

$$\begin{pmatrix} e_i^k \\ n_i^k \\ u_i^k \end{pmatrix} = R_x^u \begin{pmatrix} X^k - X_i \\ Y^k - Y_i \\ Z^k - Z_i \end{pmatrix} \quad with \tag{23}$$

The transformed variance–covariance matrix (Cuu) described in the local Eulerian system is determined considering the variance propagation law:

$$R_x^u = \begin{bmatrix} -\sin\lambda & \cos\lambda & 0 \\ -\sin\phi\cos\lambda & -\sin\phi\sin\lambda & \cos\phi \\ \cos\phi\cos\lambda & \cos\phi\sin\lambda & \sin\phi \end{bmatrix} \tag{24}$$

The number of satellites is not the unique factor in deciding when it is the best time to undertake field measurements. The geometry of the constellation is an important factor too. This factor is determined with the DOP factors. The DOP depends on the diagonal elements of the cofactor matrix and it can be decomposed into the following components:

$$C_{uu} = R_x^u C_{XX} R_x^{uT} \tag{25}$$

DOPs can be designed: it is not necessary to take any measurements, because the cofactor matrix can be obtained "a priori".

This operation called "planning", only requires knowing the receiver position and the almanac of the ephemerides used to determine the satellite's position. DOPs are like an instantaneous picture of the constellation. High values (for example GDOP > 6) are inadvisable to reach a good precision.

Azimuth A, elevation E and distance d of the different satellites can be estimated in order to plan the measurement session. Each element is determined by the following equations:

$$\begin{aligned} GDOP &= \sqrt{\sigma_e^2 + \sigma_n^2 + \sigma_u^2 + \sigma_t^2} \\ PDOP &= \sqrt{\sigma_e^2 + \sigma_n^2 + \sigma_u^2} \\ HDOP &= \sqrt{\sigma_e^2 + \sigma_n^2} \\ VDOP &= \sqrt{\sigma_u^2} \\ TDOP &= \sqrt{\sigma_t^2} \end{aligned} \tag{26}$$

An example is given in Appendix A to illustrate the various aspects discussed in this chapter.

Appendix A

DOP and satellite visibility

Estimate GDOP, PDOP, HDOP, VDOP, TDOP factors at P with coordinates: φ=45°3'48", λ=7°39'41", h=0 m, referenced in WGS84 (a=6378137 e2=0.006694379990) ellipsoid, on 2012/04/06 at 8:53:59 am. Moreover Calculate the "satellite visibility" at P. The ECEF coordinates of the visible satellites are known [8]:

Table 2. Satellite's coordinates in ECEF

sat	X	Y	Z
1	22504974.806	13900127.123	-2557240.727
2	-3760396.280	-17947593.853	19494169.070
4	9355256.428	-12616043.006	21189549.365
7	23959436.524	5078878.903	-10562274.680
10	10228692.060	-19322124.315	14550804.347
13	23867142.480	-3892848.382	10941892.224
17	21493427.163	-15051899.636	3348924.156
20	14198354.868	13792955.212	17579451.054
23	18493109.722	4172695.812	18776775.463
31	-8106932.299	12484531.565	22195338.169
32	8363810.808	21755378.568	13378858.106

In the following, the solution is provided.

The geographical coordinates are transformed to a DEG system:

$$E = atn\frac{u^{j}}{\sqrt{n^{j2} + e^{j2}}}$$

$$A = atn\frac{e^{j}}{n^{k}}$$

$$d = \sqrt{e^{j2} + n^{j2} + u^{j2}}$$

from which the geocentric coordinates of P are obtained.

$\varphi = 45.07333333\,°$ $\lambda = 7.791388889\,°$

P will be the origin of the Eulerian local system. A matrix is calculated in the ECEF system considering (11), with the following values:

Table 3. Elements of A matrix

	-0.7677	-0.5662	0.3001	-1
	0.3262	0.7350	-0.5944	-1
	-0.2235	0.6050	-0.7642	-1
	-0.7786	-0.1789	0.6015	-1
	-0.2497	0.8644	-0.4364	-1
A=	-0.9268	0.2148	-0.3082	-1
	-0.7352	0.6761	0.0494	-1

	-0.4637	-0.6290	-0.6240	-1
	-0.6896	-0.1756	-0.7026	-1
	0.5081	-0.4800	-0.7151	-1
	-0.1672	-0.9090	-0.3819	-1

From which the cofactor matrix as (13) and (18) are obtainable using the weight matrix equal to the identity (P=I)

Table 4. Elements of cofactor matrix

	0.6861	0.0092	0.4168	-0.3953
$Q_{xx}=N^{-1}=A^TA=$	0.0092	0.2541	0.0463	-0.0149
	0.4168	0.0463	0.7464	-0.3999
	-0.3953	-0.0149	-0.3999	0.3705

The rotation matrix in P is defined by (23) and it is equal to:

Table 5. Elements of rotation matrix

	-0.1333	0.9911	0.0000
$N=\frac{a}{\sqrt{1-e^2\sin^2\phi}}=6388862.01\ m$ $X=(N+h)\cos\phi\cos\lambda=4472328.363\ m$ = $Y=(N+h)\cos\phi\sin\lambda=601613.841\ m$ $Z=[N(1-e^2)+h\sin\phi=4492322.547\ m$	-0.7016	-0.0944	0.7063
	0.7000	0.0942	0.7079

The cofactor matrix is now calculated in a Eulerian coordinate system considering (24), applying this equation only to the position components (grey elements in the Qxx matrix) and not to the time element:

Table 6. Elements of cofactor matrix in Eulerian coordinate system

	0.2594	0.0273	-0.0409
Quu =R u=	0.0273	0.2943	0.0319
	-0.0409	0.0319	1.1329

and using (25), DOP factors can be determined:

$$R_x^u Q_{xx} R_x^{uT}$$

Before estimating the azimuth, elevation and distances, it is necessary to estimate the satellite position in an Eulerian system.

The satellite visibility can be estimated using (26). The final results are as follows.

Table 7. Satellite visibility

sat	e	n	u	ele (°)	azi (°)	dist (m)
1	10775718.505	-18885463.599	8885172.533	**22.227**	**150.292**	**23488787.44**
2	-17286050.680	18122569.546	3109852.920	**7.078**	**316.353**	**25237001.81**
4	-13750650.216	9615363.143	13993256.686	**39.827**	**304.964**	**21848268.18**
7	1839308.792	-24727521.634	3405994.949	**7.821**	**175.746**	**25028667.26**
10	-20513313.116	4946360.859	9273705.868	**23.725**	**283.557**	**23049167.47**
13	-7040025.788	-8627144.285	17719096.273	**57.854**	**219.216**	**20927397.22**
17	-17783003.382	-11271791.398	9631674.716	**24.582**	**237.631**	**23152918.79**
20	11776928.073	1175359.814	17314805.970	**55.646**	**84.301**	**20973316.65**
23	1669976.573	-84099.847	20262888.169	**85.283**	**92.883**	**20331761.64**
31	13453888.242	20207851.250	4844990.565	**11.286**	**33.655**	**24755570.98**
32	20446124.731	1550262.123	11006751.692	**28.226**	**85.664**	**23272213.30**

References

1. Borre, K, & Strang, G. Linear algebra, geodesy and GPS, Wellesley (USA), Wellesley-Cambridge Press, (1997).
2. Cina, A. GPS. Principi, modalitŕ e tecniche di posizionamento, Torino, CELID, (2000).
3. Hoffmann Wellenhof B Lichtenegger H., Wasle E., GNSS Global Navigation Satellite System. GPS, GLONASS, Galileo & More, Austria, Springer Wien NewYork, (2008).
4. Leick, A. GPS satellite surveying. Wiley (3th edition), (2003).
5. Global Navigation Satellite System GLONASSInterface Control Document. Navigation Radiosignal in Band L1, L2 (edition 5.1). Moscow, (2008).
6. GPS website: http://www.gps.gov/technical/ps/2008-SPS-performance-standard.pdf (accessed 25/08/2012).
7. Bern university: http://gaussgge.unb.ca/GLONASS.ICD.pdf (accessed 25/08/2012)
8. International GNSS service http://igscbjpl.nasa.gov/components/prods_cb.html (accessed 25/08/2012)
9. Blewitt, Geoffrey. "Basics of the GPS Technique: Observation Equations". http:// www.nbmg.unr.edu/staff/pdfs/Blewitt%20Basics%20of%20GPS.pdf. (accessed31/7/2012).
10. European Union (2010): "European GNSS (Galileo) Open Service Signal In Space In- terface Control Document": http://ec.europa.eu/enterprise/policies/satnav/galileo/open-service/index_en.htm (accessed 31/7/2012)
11. 05 European reference system. http://www.euref.eu/symposia/book2003/P-05-Kaniuth.pdf (accessed 25/08/2012

[12] International GNSS service: RemondiB., Spofford P.: "The National Geodetic Survey Standard GPS Format SP3". http://igscb.jpl.nasa.gov/igscb/data/format/sp3_docu.txt. (accessed 25/08/2012)

[13] GPS Website: Global Positioning System Directorate System Engineering & Integration Interface Specification IS-GPS 200. 21 Sept. 2011. http://www.gps.gov/technical/icwg/#is-gps-200 (accessed 31/7/2012).

Chapter-4

Analysis of Ambiguity False Fixing within a GBAS

1. Introduction

Since the first appearance of satellite positioning systems, GNSS positioning has become a standard and common practice. It is available in most parts of the territory of many nations. So, it is necessary to focus attention not on the feasibility of the positioning itself but on its quality control. In particular, we focus attention on the NRTK (Network Real Time Kinematic) positioning of a geodetic receiver into a network of permanent stations, the CORS (Continuous Operating Reference Stations) network. The goal of the survey is to obtain a centimetre accuracy, which is usually achieved after a correct fixing of the ambiguity phase. For various reasons, however, it may be possible that the fixing of the integer ambiguities in the receiver can be unreliable. Although the number of false fixes of the ambiguity is limited, it is necessary to minimize these events, or to know some available control parameters needed to alert the user when this eventuality is likely.

The causes of a false fix are mainly a result of four factors:

1. inaccurate differential corrections provided by the network;
2. unreliable data transmission or high latencies;
3. disturbed or partially occluded measurement site;
4. electronic or algorithmic problems in the receiver.

Not all of these parameters are controllable or modifiable; some of them can be controlled by the network manager, while others can be controlled by the rover receiver user.

For these purposes, numerous NRTK experiments were performed over many days, using some receivers connected to an antenna settled on an undisturbed site, equidistant from the CORSs stations. Note the position of this station is considered as a 'false fix' (hereinafter also called FF) for all the real-time positioning of the rover receiver greater than a tolerance threshold. Later, we analyse the time series of some parameters available in real time (both on the user and network server side) during the NRTK positioning, in order to predict with a significantly high probability what the "most probable" situation for a false fix would be.

1GBAS is an acronym for Ground Based Augmentation System. For more details see [9]

The analysis provides an estimator of the ambiguity fixing validity, useful for providing an intuitive and synthetic information for a user during the survey, similar to a "traffic light", and depending on the level of reliability of the solution. Similarly, it is possible to imagine the same architecture transferred to the following inverse logic: if, in fact, the rover receiver transmits the raw data to the network, it is possible to understand more easily if the positions obtained are a false fix or not, and it is therefore possible to notify the user in real time during the survey. In any case, it is then possible to verify whether the false fixes are a result of some quality parameters (and in what percentage) depending on the network in real time, or if they are found in the minimal information that the receiver is able to transmit to the network, contained primarily in the NMEA (National Marine Electronics Association) messages.

2. Network positioning concepts

Carrier-phase differential positioning has known an enormous evolution owing to phase ambiguity fixing method named 'On the Fly' Ambiguity Resolution since 1992 (for example see [7]). Using this technique, a cycle slip recovery, also for moving receivers, was not prob- lematic, but positioning problems, when distances between master and rover exceed 10–15 km, had not yet been resolved. For this reason, at the end of the 1990s, the Network Real Time Kinematic (NRTK) or the Ground-Based Augmentation System (GBAS) was designed, as in [4].

First, to understand the network positioning concept, it is necessary to keep in mind some concepts about differential positioning. To do this, it is possible to write the carrier phase equation in units of length:

$$\phi_k^p(i) = \rho_k^p - cdT_k + cdt^p - \alpha_i I_k^p + T_k^p + Mi_k^p + E_k^p + \lambda_i Ni_k^p + \varepsilon_k^p \tag{1}$$

In this equation, ϕ *p(i)* represents the carrier-phase measurement in units of length between the satellite p and the receiver k on the i-th frequency. On the right-hand side of the equation, in addition to the geometric range ρ^p, it is possible to find the biases related to receiver and satellite clocks multiplied by the speed of light (cdT_k and cdt^p), the ionospheric propagation delay $\alpha_i I_k^p$ (with a known coefficient $\alpha = f_1^2/f_i^2$ that depends on the i-th frequency), the tropospheric propagation delay T_k^p, the multipath error Mi_k^p, the ephemeris error E_k^p, the carrier-phase ambiguity multiply by the wavelength $\lambda_i N i_k^p$ and, finally, the random errors ε_k^p. Single differences can be written considering two receivers (h and k). Neglecting a multipath error, that depends only on the rover site and therefore cannot be easily modelled, it is possible to write:

$$\phi_k^p(i) = \rho_k^p - cdT_k + cdt^p - \alpha_i I_k^p + T_k^p + Mi_k^p + E_k^p + \lambda_i Ni_k^p + \varepsilon_k^p \tag{2}$$

Focusing the attention to the orbit error E_{hk}^p, it is possible to affirm that this term can be neglected for sub-regional networks, using the IGS ultra-rapid products [10], while they can also be modelled as happens into the GNSMART™ software [11].

After that, double differences equations can be written considering two receivers (h and k) and two satellites (p and q). By subtracting the single difference calculated for the satellite q from the one calculated for the satellite p, it is possible to obtain the double differences equation, neglecting the random errors contribution:

$$\phi_{hk}^{pq}(i) \text{ that means } \phi_{hk}^{p}(i) - \phi_{hk}^{q}(i) = \rho_{hk}^{pq} + \lambda_i Ni_{hk}^{pq} - \alpha_i I_{hk}^{pq} + T_{hk}^{pq} + E_{hk}^{pq} \tag{3}$$

When the distance between the two receivers is lower than 10 km, the atmospheric propagation delays and the ephemeris errors are irrelevant, allowing centimetric level of accuracy. Over this distance, these errors increase and cannot be neglected. Otherwise, these errors are spatially correlated and can be spatially modelled (as possible to see in [5]). However, to be able to predict and use these biases in real time, three conditions must be satisfied: the knowledge of a centimetric accuracy of the master positions; a control centre able to process data from all the stations in real-time; and the continuous carrier-phase ambiguity fixing, even when interstation distances reach 80–100 km. (This last condition in the control centre network software is called "common ambiguity level fixing"). This concept is equal to bring to the left- hand side of (3), among the known terms, the first two terms on the right-hand side, i.e.:

$$\phi_{hk}^{pq}(i) - \rho_{hk}^{pq} - \lambda_i Ni_{hk}^{pq} = \alpha_i I_{hk}^{pq} + T_{hk}^{pq} + E_{hk}^{pq} \tag{4}$$

If the software uses the first differences, or undifferenced phase equations, in the first case, it is essential to calculate and model the clock errors of the receivers and, in the second case, also those of the satellites, in addition to a series of secondary biases that are usually not considered in differencing techniques. After the phase ambiguity fixing for all the network stations, it is possible to use the iono-free and geometry-free combinations to separate the ionospheric delay by the geometric one. In this way, having these biases for all the stations, it is possible to model the residual ionospheric and tropospheric biases and the ephemeris error, not only between stations h and k, but also among all the reference stations of the network. When these errors are modelled spatially, they can be interpolated and broadcast to any rover receiver.

The differential corrections will no longer be broadcast by a single master station, but by the network centre. It is not always necessary for the network software to send the interpolated biases to the rover receiver. Once the dispersive ($\alpha_i I_{hk}^{pq}$) and non-dispersive ($T_{hk}^{pq} + E_{hk}^{pq}$) biases for each station and each satellite have been estimated, it is possible to follow three paths:

- ❖ to transmit data from one network station (called the master station), together with the first differences of a few stations, called "auxiliary" stations, close to the rover receiver (Master Auxiliary Concepts - MAC mode);
- ❖ to model these biases within the area of the network and transmit the parameters of this model (Flächen Korrektur Parameter - FKP mode);
- ❖ to interpolate these data close to the rover receiver and build some synthetic data for this receiver as if they were generated by a station very close to the receiver. (Virtual Reference Station - VRS® mode).

In MAC mode, the receiver's task is to use data provided by a master station and the first differences of some auxiliary stations. It can then perform a multi-base position and the biases can be interpolated as necessary. The amount of data and calculations that the rover receiver can do is considerable. On the other hand, the multi-base positioning can be much more accurate than other methods. Finally, the user chooses the best way to use the data in a subnet. The NRTK network must only carry out the ambiguity fixing and transmit MAC corrections. Inside this cell the same correction is identical for all rover receivers and could also be transmitted by radio.

In the FKP method, the network software models the geometric parameters and dispersive biases over a wide area. Subsequently, the data derived from a master station and the param- eters of this model are transmitted to the rover. The "one-way" approach for this type of correction could also be used in this case. Compared to the MAC method, the network software must also model the biases in the area of the network correctly.

Considering the VRS® positioning, the task of the network software is not only to model, but also to interpolate these biases, close to the position of the receiver that "sees" these corrections as if they came from a master station that actually exists. The task of the network software increases and decreases that of the receiver (as possible to see [1]).

In all three cases, it is necessary to control the quality of the positioning, both from the side of what provides the network, and from how the generic network product is used by the rover receiver. The quality of the positioning is a parameter that must be monitored in real time to avoid a wrong phase ambiguity fixing happening (also called "false fixing": FF). Often this is a result of internal problems of the network software or, more often, the poor quality of the environment in which the rover receiver works.

3. Strategy for the control of the NRTK fixing

To achieve this control, a tool was designed that, starting from the data available in real time by a user connected to an NRTK positioning service, can identify with a certain probability threshold (the calibration on this tool depending on the type of receiver available) the effective presence, or the possibility of a false fixing in the position. The input data of this instrument (hereinafter referred to as the "FF Estimator") will be the real-time data available from the NMEA message, extractable from the rover receiver (number of satellites in view, Diluition Of Precision – DOP - index, delay of the differential correction), and those which may be obtain- able by the network software (typically, the tropospheric and ionospheric delay, the number of satellites fixed by the network, the network quality fixing).

Figure 1. Algorithm procedure

The FF estimator will be composed of a neural network, trained a priori with some datasets, and will have, as a single output, the probability that the current fixing is a false fixing of the ambiguity of phase. In this regard, it will be necessary to identify, among all available data in real time, those most sensitive both to the deterioration of the accuracy and the presence of the false fixing. These parameters will be those used to calibrate ("train") the neural network. Downstream of the estimator, a representative index is provided as an output of the algorithm (shown schematically in Figure 1 by a traffic light), of the quality of the fixing, possibly allowing the user to re-initialize the measurement session. The change over time of these probabilities is useful to forecast an incorrect positioning (not always considered as a false fix) before it actually occurs.

4. Notes on neural network

Artificial neural networks are informatic tools that simulate the connections of neurons in the human brain, and are used to make decisions, approximate a function, or classify data. In our case, the function or the decision is to assess the probability of a false fixing, and provide as output an indicator simulating the behaviour of a traffic light, allowing the user to have visual and intuitive information about the quality of the coordinates of the NRTK survey, available in real time.

Neural networks are made up of n nodes, called artificial neurons, connected to each other, which, in turn, are ordered in layers. There is one input layer consisting of a limited number of neurons, one output layer (in our case constituted by a single neuron that represents the decision on the false fixing), and one or more intermediate layers. Each neuron communicates with one or more neurons in a subsequent layer. The first layer receives information from the input and returns it in the next layer through interconnections. The connections between one layer and the next do not transmit the output values derived from the previous layer directly, but transmit weighted values through weights wij that are assigned to each of these connections. These weights are the unknowns of the problem; to calculate them it is necessary to "train" the network. In other words, it is necessary to perform a training phase which is based on providing, as input, a set of data for which you know the result in output. On each neuron of a layer the weighed contributions of neurons of the previous layers and a generally constant bias value are summed into the input. A function is applied at this sum and the result is the output value from that neuron (for example see [6]).

As mentioned above, it is necessary to provide a training phase of the neural network; there are essentially three ways: an *unsupervised learning*, one called reinforcement learning and one *supervised learning*. The first method (unsupervised) allows to change the weights considering only the input data, trying to evaluate in the data a certain number of clusters with similar characteristics. The second method (reinforcement) follows a supervised strategy but it can be adjusted depending on the response of the external application of the parameters obtained. The last mentioned mode (supervised) is used when, as in the present case, a dataset is available (called the *training set*) for a training phase. The algorithm of supervised learning implemented is the so-called "backpropagation", which uses both the data provided by the input and the known results in output, to modify the weights so as to minimize the prediction error called *network performance* (goal function). In the cases in this work, we need to minimize the mean square error:

$$E_2 = \frac{1}{n}\sum_{i=1}^{n}(T_i - O_i)^2 \tag{5}$$

where O_i is the neuron output of the last level and T_i is the known value that we want. This process is iterative: the first iteration of the associated weights between neurons of two adjacent layers have values small and random, which are then updated at subsequent iterations. There are two ways to implement the training: the incremental mode; and the batch mode. In the first, the weights are updated after each input is applied to the network. In the second, all the inputs in the training set are applied to the network before the weights are updated (for example see [8]). To do this it is possible to use the gradient of the network performance compared to the weights, or the Jacobian of the network errors (T_i – O_i) with respect to the weights. Gradient or Jacobian are calculated using the "backpropagation" technique, which gets the values of the weights by using the chain rule of calculus. The training algorithm used in these application is the Matlab® routine "trainlm", based on Levenberg-Marquardt optimization. The training phase occurs in two distinct phases, called the forward pass and the backward pass, respectively. The first phase allows both to derive the estimation error between the input data (applied to the input nodes) and the results obtained in output (which are known), and to propagate this error in the reverse direction, while in the second phase, the weights are modified in order to minimize the difference between the real and the known, desired output.

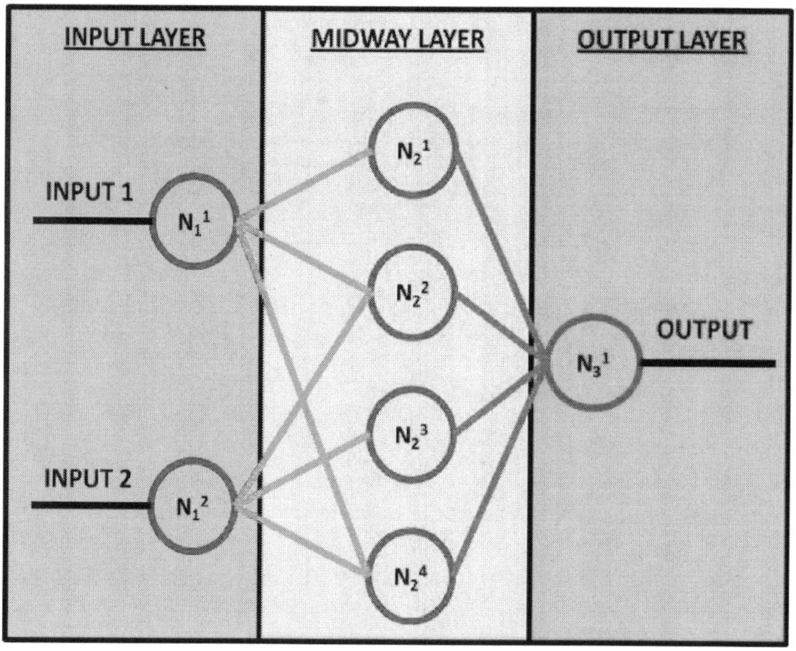

Figure 2. Oversimplification of a neural network

In Figure 2, the generic neuron k of the j-th layer is indicated by N_j^k . Neurons with subscript 1 are the input neurons, whose number must be greater than or equal to the number of input. The input value I, for a generic neuron j of the k-th layer which is not of the first level, is therefore given by:

$$I_j^k = b^k + \sum_{h=1}^{l} w_h \tag{6}$$

where l represents the number of neurons of the previous layer, w_h the weights of these neurons, and b^k the value of the bias. So for each neuron that is not at the first level, the inputs of the layer above, weighted with weights w , are summed, adding a further bias b . The neurons process the input through a certain function called transfer function: of all the transfer functions, the Hard Limit, the Linear and Log Sigmoid are the best known (Figure 3). In addition to the transition function, each neuron is characterized by a threshold value. The threshold is the minimum value that must be present in input that allows the neuron to send something in output and is, therefore, active. Beyond this threshold value the neuron transmits a value to the next layer that is the result of the application of the transfer function.

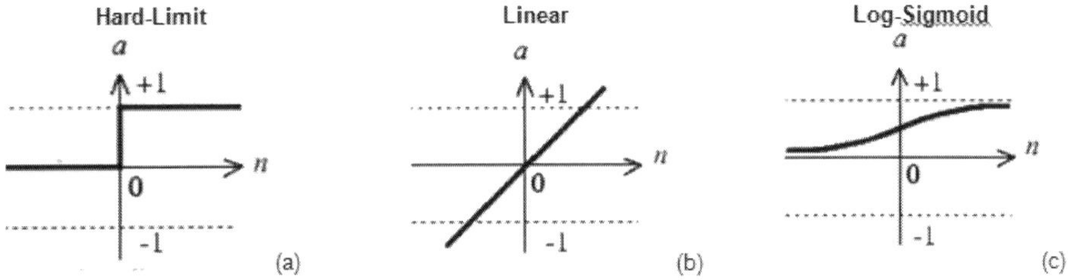

Figure 3. Different types of Transfer Function

Generally, between one level and the next, all the neurons of the first level are connected with all the neurons of the second. It should be emphasized that there are neural networks in which there are also connections between neurons of the same level, but this does not happen in the case presented in this chapter. After the training phase, the effectiveness of the network is tested on a new set of data, called the test set. This set of data must be constituted by values of input and output never seen by the network. If the results offered by the network are close to the actual values, then the network can be considered valid. In these cases, it is said that the network is effective in generalization.

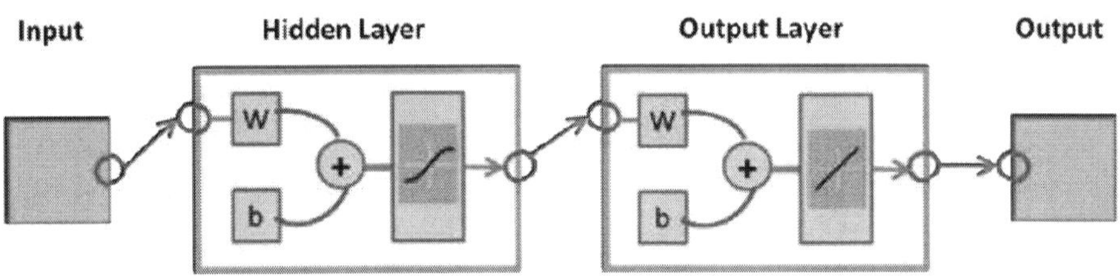

Figure 4. Neural network diagram

The neural network used in these experiments is shown schematically in Figure 4. The input consists of three vectors which are, for all epochs, the values of the HDOP

(Horizontal Diluition Of Precision) index, the latency of the correction, and the number of fix satellites seen by the rover. The training set is constituted, for every epoch, by a Boolean vector consisting of 1, in the case of a false fix, and 0 otherwise. The hidden layer consists of three neurons. The transfer function in the hidden layer is the Log-Sigmoid, while the layer is linear for output. The training function used is (always) the Levenberg-Marquardt backpropagation Algorithm.

The dataset for training the network was divided as follows: 50% training; 30% validation; 20% testing. As mentioned, the control of what causes a FF is executed simultaneously in two respects: the verification of what is happening on the network at the time of the FF; and the control of what the user can see on the rover receiver. The quality parameters, reported and considered significant to identify the reason for the FF, are extended for a period before and after the false fixing, established at five-minute intervals (300 s). Almost always this period, especially before the false fixing, is sufficient to understand what is injuring the positioning.

5. Analysis of the quality parameters

5.1. Quality parameters of the NRTK network

Before testing the developed algorithm, it was necessary to choose the network architecture to support the measures. We chose the Geo++ GNSMART ™2 http://www.geopp.de/ index.php?sprachauswahl=en&bereich=0&kategorie=31&artikel=35 software as the network software of permanent stations because it allows the user to extract information on the real- time calculation of the network: it is possible to have the state vector of the network calculation as an output, every 15 seconds. This information is transmitted to the server, in an ASCII format, and links the age of the false fix with a network event that may cause estimation problems for the rover receiver. Within the software GNSMART (version 1.4.0) a network of five stations was therefore set up, settled around the measurement site and having an inter- station distance of about 50 km. This type of configuration allows the simulatation of meshes of today's typical positioning services, and was therefore considered to be representative of the tests performed.

The permanent stations involved in these tests are shown in Figure 5 and detailed in Table 1:

Table 1. Permanent stations involved in the NRTK test

Station ID	City	Receiver	Antenna	Distance to VERC (rover)
ALES	Alessandria	LEICA GRX1200+GNSS	LEIAR25.R3	48 km
BIEL	Biella	LEICA GRX1200+GNSS	LEIAR25.R3	39 km
CRES	Crescentino	LEICA GRX1200+GNSS	LEIAR25.R3	29 km
LENT	Lenta	TPS NETG3	TPSCR.G3	27 km
VIGE	Vigevano	TPS ODYSSEY_E	TPSCR3_GGD	35 km

2 For more details visit the webpage:

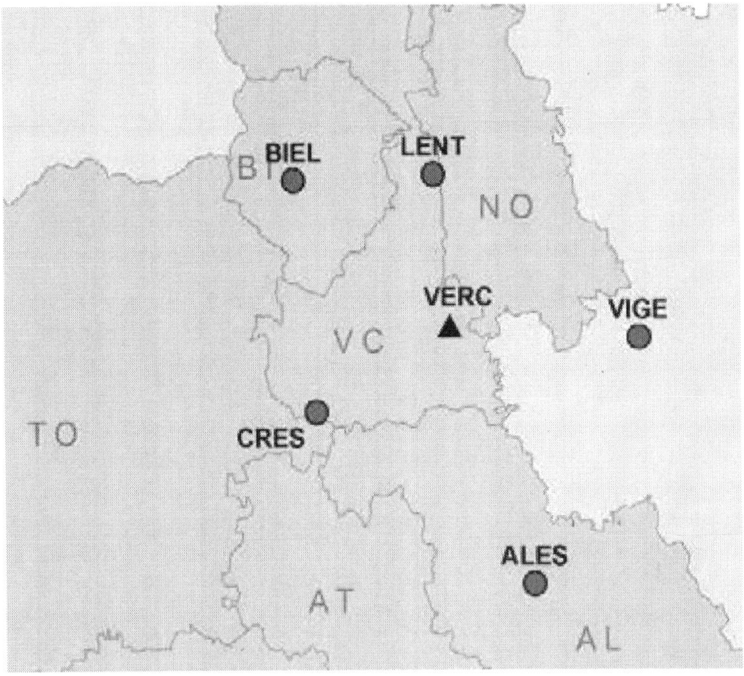

Figure 5. Permanent stations used in the NRTK test

Figure 6. Pillar and antenna used for the NRTK test (VERC station)

The reference coordinates of the stations were obtained by adjusting 15 days of data with the Bernese GPS software (v. 5.0), in the reference system IGS05, constraining the coordinates with the 5 closer IGS (International GNSS Service) stations.

As mentioned, the control of what causes a FF is executed simultaneously in two respects: the verification of what is happening on the network at the time of the FF; and the control of what the user can see on the rover receiver. Therefore, by combining the network information with the information of the receiver, a reliable decision is more likely to be reached, regarding the measurement period. The association has been possible thanks to the GPS time scale in both files: the one produced by the receiver in NMEA message; and the one produced by the network. Thereafter, for predicting the quality of the NRTK positioning, a software was developed that does not yet use the information derived from the network, but is already suitable to process these parameters. Before using any software that automatically predicts a quality index of positioning, it was necessary to analyse which parameters are more sensitive to the fixing degradation when observing the network status parameters. Since the state vector, which is obtained as output from the network, is complete, it should be noted that much of the information is not essential for checking the quality of the network data. It was therefore necessary to make a decision. The following are considered to be the only useful parameters:

- Number of GPS and GLONASS (GLObal'naya NAvigatsionnaya Sputnikovaya Sistema) satellites seen by the network stations for each age. In particular, it was decided to plot the minimum number among all satellites viewed from the stations of the network.
- Number of GPS and GLONASS satellites fixed by the network for each age.
- rms (root mean square) of the carrier phases compensation obtained in the current epoch,
- expressed in metres (σ_{ep}).
- The value of the ionospheric delay expressed in ppm along the vertical direction. This index, that is a characteristic value of the GNSMART software, is called I95 and is indicative of the error obtained from the global estimation of the ionosphere (for example see [3]):

Table 2. Values of I_{95}

First Order Effect of Ionosphere/Distance Dependency	I_{95} [ppm]
less	0 ÷ 2
small	2 ÷ 4
large	4 ÷ 8
severe	"/> 8

- The variation of the tropospheric delay (dry component) for each station in the network (expressed in percentage and derived by the modified Hopfield model).

5.2. The quality parameters available at the rover receiver

The station, which for several days simulated a rover receiver, was a receiver placed in the Geomatics Laboratory at Vercelli; this receiver was connected with a geodetic antenna settled on the roof of the same laboratory (Figure 6). The coordinates of that station were calculated in the same network frame of reference system, through a post-processing approach of 24 hours of data with the nearest station (LENT).

For each network product used (FKP, MAC, VRS) (Table 3) and for each geodetic instrument, RTK positioning with 24 hours of session length were performed, with an acquisition rate of 1 s, in order to obtain results that were independent of the constellation view from the GNSS receiver. This chapter will show the results of a receiver which we believe has shown repre- sentative average behaviour with respect to all the receivers used. For each network product, both the raw data and the real time positions were saved: the last one is obtained through the GGA field of the NMEA message, transmitted via a serial port to a computer close to the receiver in the acquisition.

Table 3. Network products considered in the NRTK test

Type of product used by receiver	Type of format
Master Auxiliary Concept (MAC)	RTCM 3
Flachen Korrektur Parameter (FKP)	RTCM 2.3
Virtual Reference Station (VRS®)	RTCM 3

If the raw data of the receiver was saved mainly for the analysis of the quality of positioning in post-processing, the data obtained in real time is particularly useful for studying the causes and effects of a FF.

From the NMEA message sent by the receiver we extracted the following information:

- mean square error (rms) of the position, calculated on the 30 s before the current epoch;
- rms of ellipsoidal height, also calculated 30 s before the current epoch;
- the age of differential correction broadcast;
- number of satellites received;
- HDOP (Horizontal Dilution of Precision) index.

5.3. Search of the different types of FF and its related parameters

To try and assess any FF occurred during the measurement phase, the reference coordinates were compared with those obtained in the NRTK survey by the rover receiver, considering only the "fix" coordinates, where "fix" means that the ambiguity value of the carrier phase measure- ments were fixed to the integer correct value. As already mentioned, several geodetic receiv- ers were tested, as well as different types of network products. Not all results are shown, but the idea here is to present the most significant ones. Another threshold was considered: the limit of acceptability of the three-dimensional error in 20 cm was established, in order to assess whether or not the event is a FF. This value is fairly generous and was chosen as corresponding approxi- mately to a wavelength of the carrier phase L1. Three-dimensional errors less than 20 cm do not necessarily correspond to epochs where the phase ambiguity is fixed in the correct way: FFs may exist for which the positioning errors are much lower than 20 cm. However, an ambiguity fixing that shows a three-dimensional error (i.e. the difference between the "true" position and the measured one) of more than 20 cm is definitely a false fix. Another threshold is the minimum length of the false fix (i.e. the minimum time necessary to be considered is significantly dangerous). This limit is imposed as equal to 5 seconds, because FFs with a lower length are too short to be considered deleterious. It is possible to think now not a mobile positioning, but at a NRTK "stop & go" survey. In this hypothesis, it is expected that the surveyor acquire a point for more than 5 seconds and perform the average of the coordinates obtained, thus easily identifying the measure that causes the false fixing. This option can certainly be removed or modified in the choice of the FF, and during the training of the neural network.

The following graphs show both the network and receiver parameters that can justify and provide a false fix in some way. Two vertical stripes in green and red denotes the beginning and the end of FF. For each case of FF, a series of graphs is shown, showing the values described above in a common time scale for the network and the receiver. All analysis concern the analysis of the behaviour network-receiver within 24 hours length of RTK measurements. Considering the network shown in Figure 5, during the tests with VRS®, FKP and MAC corrections, it 21 false fixes were identified with a time length between 6 s and 1136 s (corre- sponding to 18 minutes). Many of these have a similar typology, and for that reason not all the false fixes mentioned before are reported here, only the three main types. We will now examine three emblematic cases of FF.

5.3.1. Typology of FF No. 1: Latency dependent

The first typology involves the FF identified using the MAC product. It is possible to see the graphs relating to the variation of Δ3D three-dimensional error (related to the E, N and h coordinates), and the elevation error Δh, with the respective standard deviations. It was decided not to show the graphical error related to the coordinates E and N because it is insignificant if compared to those previously mentioned. It is common for the more significant error to be the altimeter one, and there isn't a preferred direction in the planimetric one.

By analysing the first FF and considering a time interval of five minutes, it is possible to see a similar trend in the latency of the correction (Figure 8). Note that the receiver does not receive data for a period of time of a few seconds. The DOP index registered by the receiver is kept fairly constant; there are, in fact, only "steps" corresponding with variations in the number of received satellites. 150 s after the start of the FF the number of these satellites decreases below the minimum available for the positioning. It is possible that one of the four satellites corrected by the network has not been available to the receiver. It is also possible to see the graph containing the value of the root mean square σ registered by the receiver. It can be seen how this magnitude does not reveal any particular problems before or during the FF, with a trend without excessive anomalies.

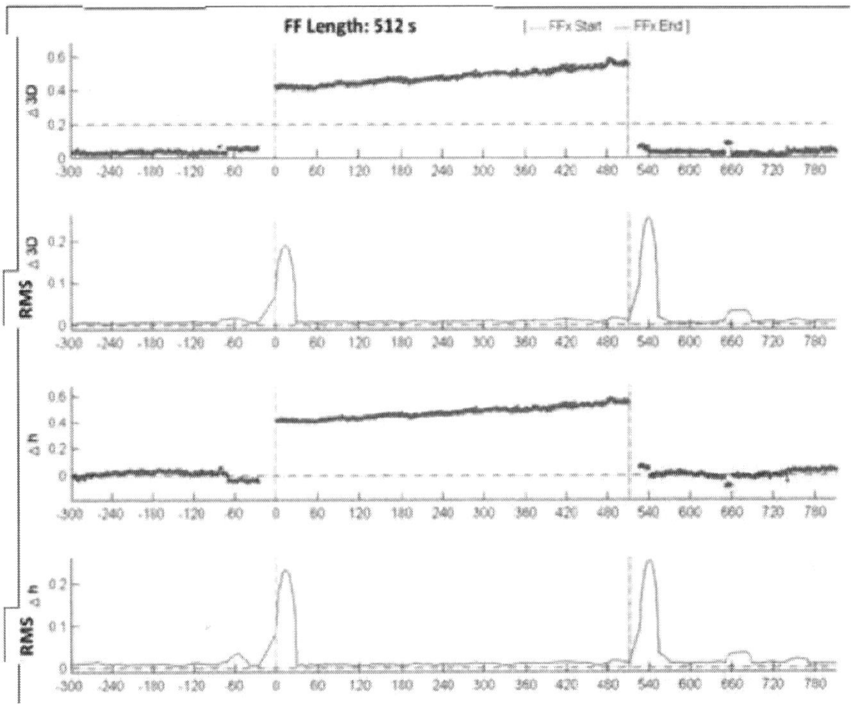

Figure 7. Residuals of coordinates for FF No. 1 (latency dependent)

The values of the ionosphere and troposphere are also considered, related to the variation of the CORS station constituting the network. Such information, which is not shown, have been extracted from the state vector of GNSMART, as described in the introductory part of the present work. In both cases, we do not obtain significant changes of the values at the beginning of the FF, while it has the presence of a peak of both ionospheric and tropospheric delay at the half of the same, which corresponds to the fixing of only three satellites by the network. The presence of the peak of the troposphere is also justified as an "attempt" by the network software to restore the correct behaviour in the biases estimation. The low number of corrected satellites compared to those seen by the receiver may be a temporary problem of the network software.

5.3.2. Typology of FF No. 2: Wrong ionospheric network estimation

This type of false fixing is unusual because there is no data before or after the FF: the first positioning with the ambiguity fixed, which occurred only 5 seconds after the switching on of the instrument, is already incorrect. The following graphs, relating to the 3D and elevation errors, show the presence of values only in correspondence to the FF error values in the order approximately of 50-60 cm, that are much higher with respect to the threshold set equal to 20 cm (Figure 9). At the beginning of the false fix, however, the rms of the position rises up to 30 cm before becoming very low.

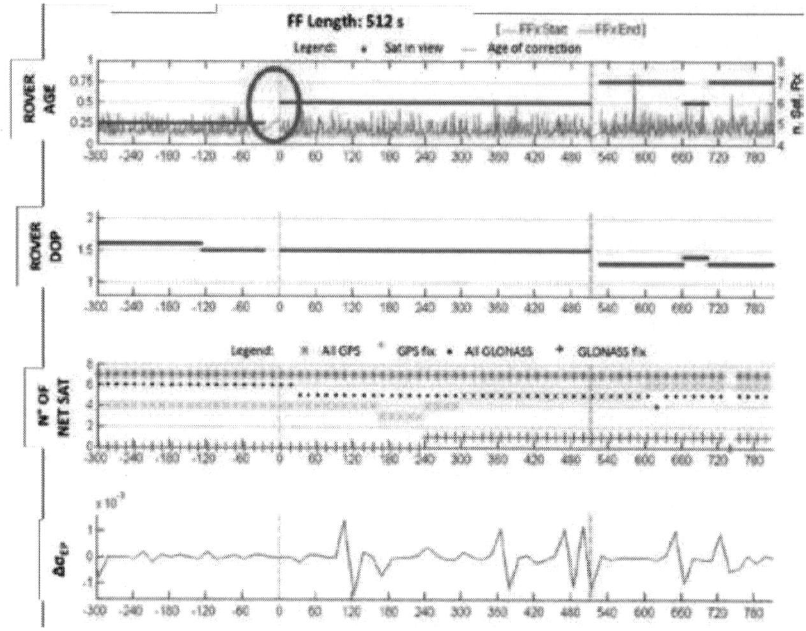

Figure 8. Quality parameters for rover and network (FF No. 1)

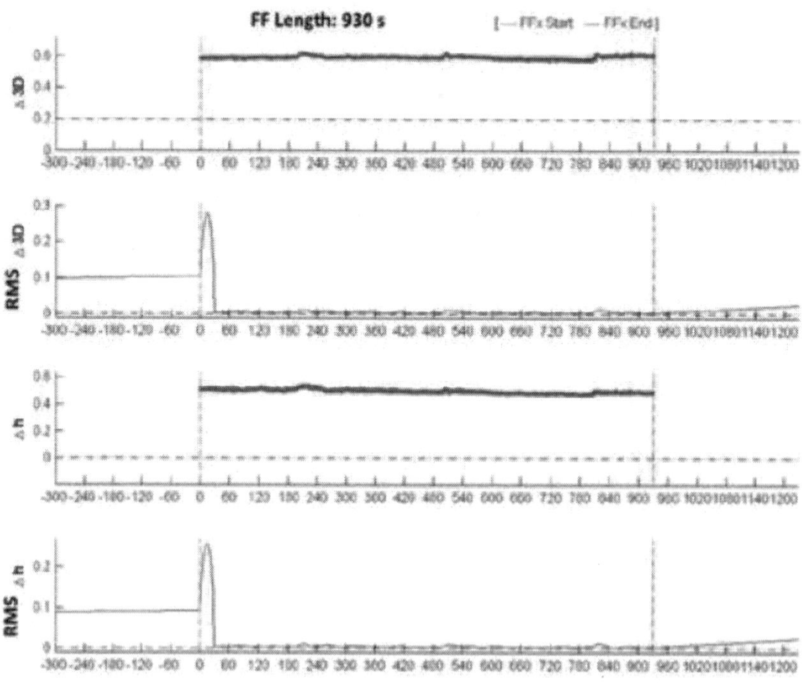

Figure 9. Residuals of coordinates for FF No. 2 (wrong ionospheric network estimation)

The graphs in Figure 10 do not show meaningful behaviour which can be attributed as the cause of the false fix. In fact, the latency of the correction is lower in magnitude than what is already found. The DOP index remains constant as the number of satellites seen by the receiver is also constant, while the number of satellites seen, and with ambiguity fixed by the network, increases (both GPS and GLONASS). Finally, the variation of the σ value is not significant during the FF, only before. There is, on the other hand, a quite significant variation in the ionospheric and tropospheric delays (the latter especially after the end of the FF, probably corresponding to the intention of restoring, using the network software, a more correct, common level of ambiguity fixing. This is shown in Figure 11).

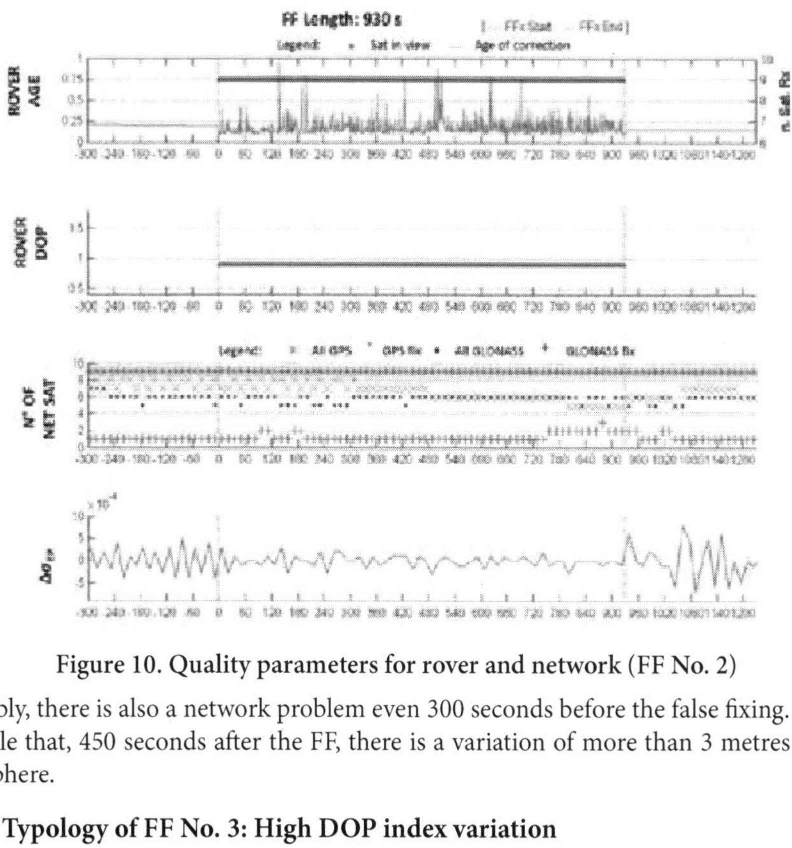

Figure 10. Quality parameters for rover and network (FF No. 2)

Probably, there is also a network problem even 300 seconds before the false fixing. It isn't possible that, 450 seconds after the FF, there is a variation of more than 3 metres in the ionosphere.

5.3.3. Typology of FF No. 3: High DOP index variation

This typology represents the false fix with the longer length, similar to type 2 with the lack of data before and after the FF, but it comes back to fixing the ambiguities around the epoch 1,300.

As seen in Figure 12, the rms of coordinates increases and exceeds the first decimeter before the FF. It is possible to know, unlike the other cases, that the value of the ionospheric delay during the FF decreases suddenly, unlike that of the tropospheric one which has a fairly constant trend (Figure 14). The DOP values, while being low at the beginning of the FF, rise after about 90 seconds, with values greater than or equal to 2, and decreasing to reasonably low values around the epoch 1300 (Figure 13).

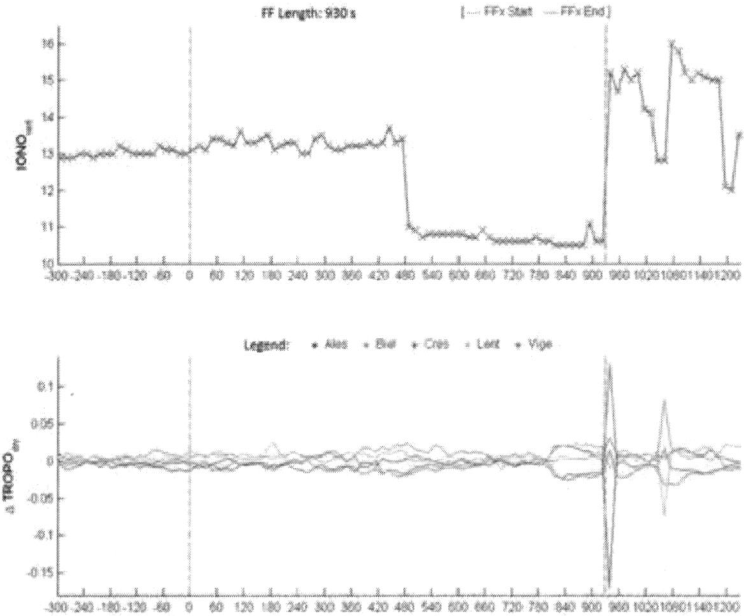

Figure 11. Atmospheric parameters estimated by the network software (FF No. 2)

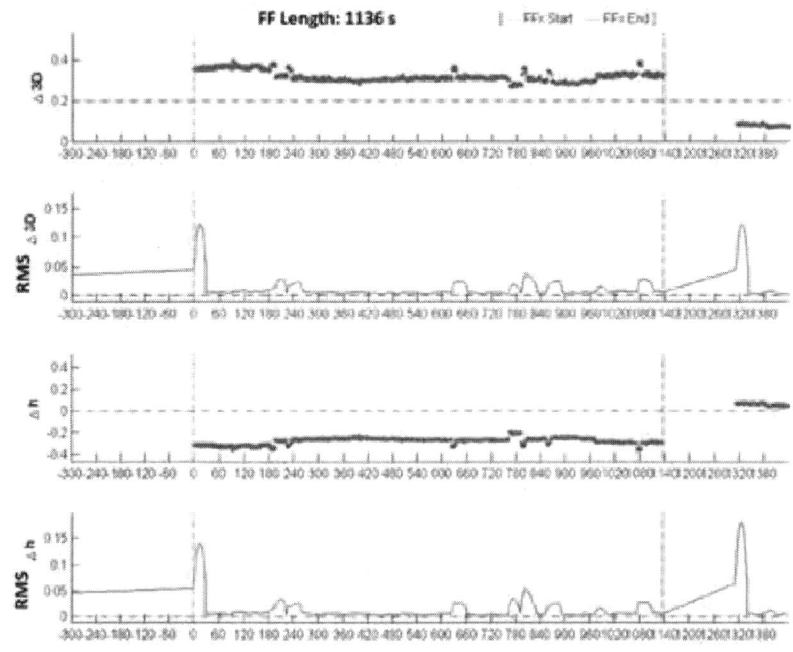

Figure 12. Residuals of coordinates for FF No. 3

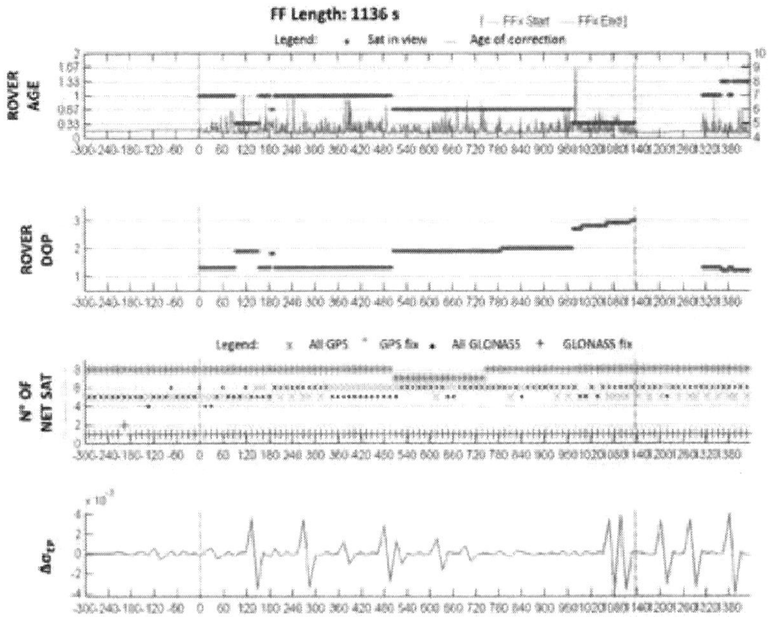

Figure 13. Quality parameters for rover and network (FF No. 3)

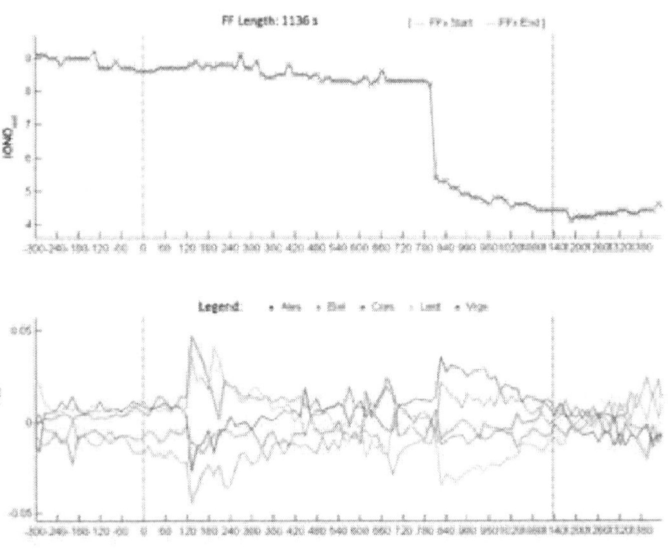

Figure 14. Atmospheric parameters estimated by the network software (FF No. 3)

Despite the existence of a FF where it is not possible to identify the main trigger, it is however possible to affirm that the significant values that enable the presence of a false fix to be identified with high probability are:

1. the latency of the correction;
2. the noise level of the σ value of the coordinates on the receiver;
3. the DOP index and then the parallel number of satellites used (parameters that can be separated);
4. the variations of the ionospheric and tropospheric biases computed by the network software.

These parameters are visible from both the receiver (points 1, 2, 3) and the network (point 4). Some of them are usually transmitted to the network software with the NMEA message, in the case of a bidirectional exchange of information. With this message, the network software is able to evaluate the state of the receiver (fixing or not).

Some analysis was carried out in real time using CORS networks with larger extension (inter- station distances of about 100 and 150 km), but the results are omitted because, with these networks, the number of FFs obtained is amplified with respect to the network presented (without changing the typology), with an inter-station distances of about 50 km [1]. This deterioration is clearly linked to a growing difficulty in the interpolation of the corrections by the network if the inter-station distances increase.

6. Implementation and results of the neural network

After the analysis of the types of false fix and the main factors that determine the occurrence of an incorrect fix, it was decided to develop a neural network to predict the wrong fix of the ambiguity phase. To do this, a "neural network" toolbox, available in the Matlab® computer program, was used. Particular attention was devoted to the training phase: it is of fundamental importance to "train" the network correctly, in such a manner that it is able to predict, after this phase, the possible false fixing of the ambiguity. Four different methods were tested, depending both on the type of the input training file (each file was composed by 3 days of observations), and the number of neurons considered in the hidden layer:

- **Net1:** the network is composed of 3 neurons; the training file derives from a real data file of 24 hours' length (containing of about 15 FF);
- **Net2:** the network is composed of 3 neurons; the training file is a file prepared ad hoc (in terms of data densification and order of FF typologies, such as FF due to latency, DOP index and number of satellites, hereafter called man-made) containing about 15 FFs, owing mainly to latency correction and sudden variation in the number of satellites;
- **Net3:** the network is composed of 10 neurons; the training file derives from a real data file of 24 hours' length (containing about 15 FFs also in this case);
- **Net4:** the network is composed of 10 neurons; the training file is a man-made file containing about 15 FFs, owing mainly to latency correction and sudden variation in the number of satellites.

To understand which types of previous network obtain the best performances, an additional day (session length of 86400 epochs) of independent measurements was used. The results obtained are shown in Figure 15.

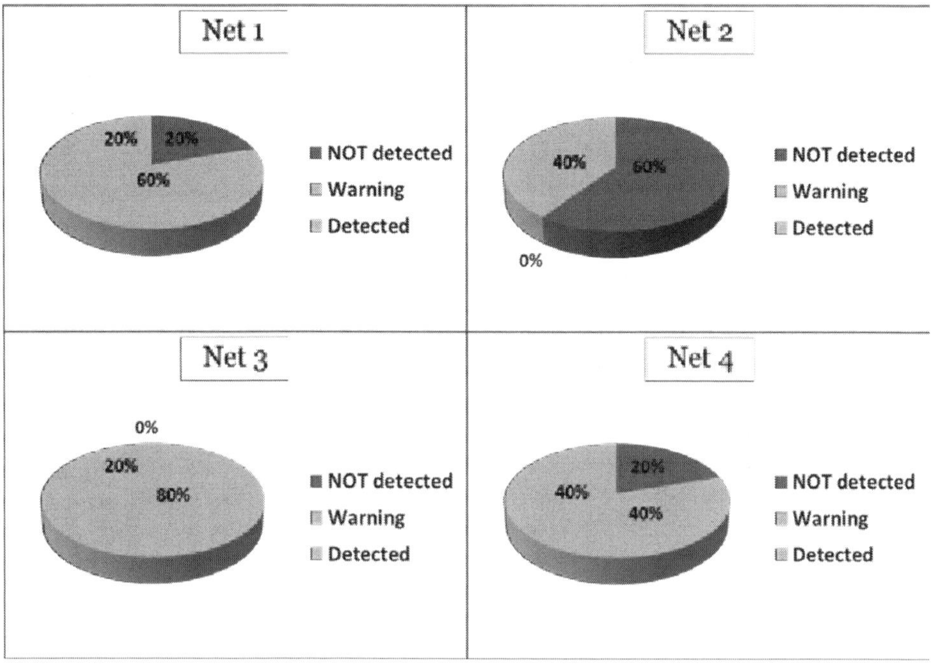

Figure 15. Neural network results (training phase): percentage of FF detected or not

It is possible to see how Net3 is the network that identifies all the FFs, declaring 20% as certain FFs while 80% are defined as possible FFs (hereinafter called also *warning*). The other networks have worse behaviour, not identifying 20% or 60% of false fixings occurring in the day used for the test phase. More specifically, and choosing a randomly FF that occurred on that day, it is possible to see that even in this case (Figure 16) Net3 predicted a FF about 5 seconds in advance. Other networks, such as Net2 or Net4, not only did not identify it, but also noticed of the event only after that it had started.

Therefore, considering Net3 as the best trained and available neural network, some tests with a session length of more than 72 hours (3 days) were carried out, in order to test the trained neural network on new data, acquired from different receivers of the main companies operating in the market. The results only show of a dual frequency and dual constellation receiver, chosen as representative (called a "medium receiver"), in terms of behaviour and quality of positioning. The information contained in the GGA field of the NMEA message was analysed, considering only the epochs with a fixed ambiguity phase. Even for these experi- ments it was considered to be a static survey with the rover instrument mentioned above, located in Vercelli (Figure 6), the CORSs network as described above (Figure 5), and only VRS® and MAC corrections.

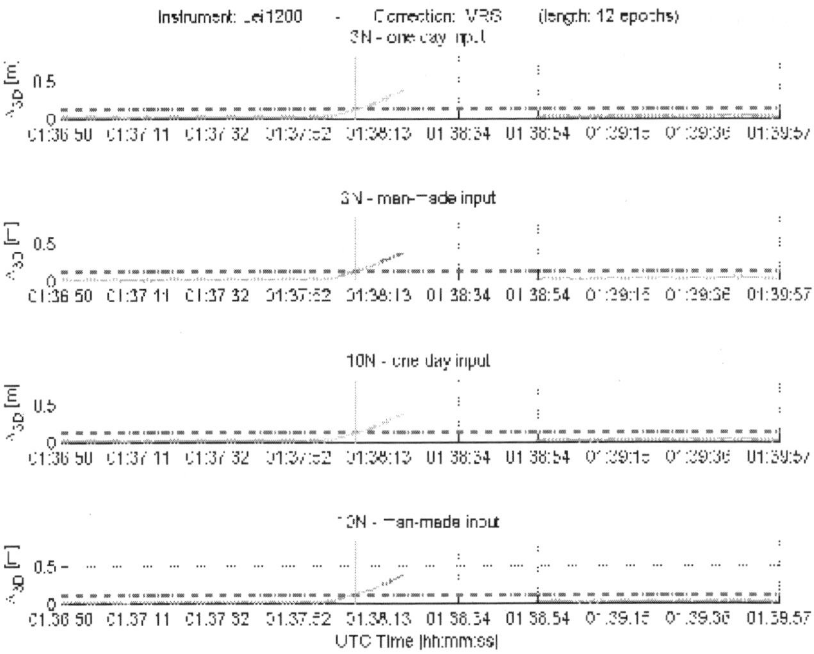

Figure 16. False fix prediction: no FF (green); warning (yellow); and FF detected (red alert)

Table 4 shows the number of FFs obtained by considering the two different corrections and three different sessions of measurement (each of which, as previously mentioned, with a duration of 72 hours). It may be noted first that, with the VRS® correction, a smaller number of false fixes is obtained generally. It should be emphasized that the experiments were not performed considering the two corrections simultaneously, but the choice of using such a big survey time window allows us to affirm that there is no dependence on the geometry of the satellite constellation, or on external factors.

Table 4. Number of false fixes / day

Day	# of False fixes	
-	**MAC correction**	**VRS© correction**
Day 1	7	1
Day 2	15	5
Day 3	18	18

By means of a special program, developed in Matlab®, it was possible to analyse all those epochs when the estimated position differs by at least 20 cm (considered as certain FF because it is a static survey) from the "correct" one. It was also possible to create some statistics results, not on the quality of positioning but on the ability to predict and identify a FF in real time. We will analyse only two cases, considered significant, and chosen from among all the sessions, analysed during the training phase: the "best case", which represents the session in which the network predicted the maximum number of FFs, indicating them as a warning or as a red alert (epochs declared as a certain FF), and the so-called "worst case", when the network predicted the minimum number of FFs.

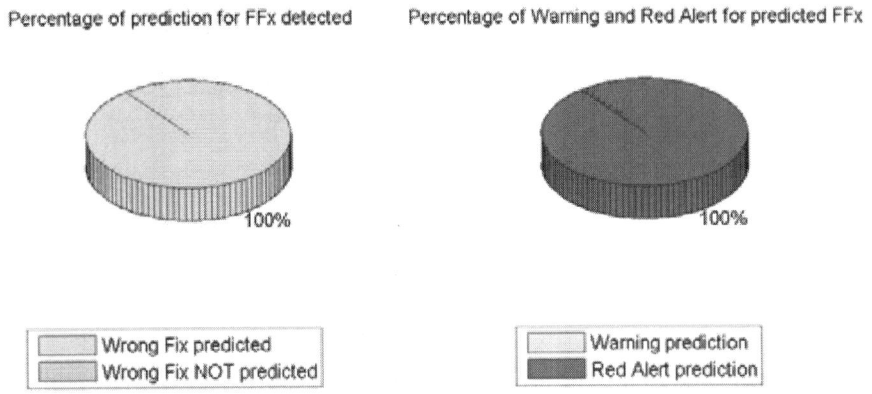

Figure 17. False fix analysis: MAC correction, "best case" (72 hours of data, Net 3)

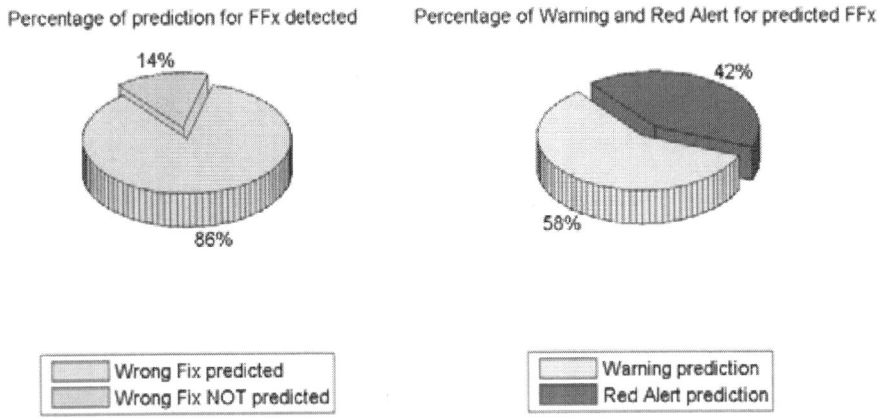

Figure 18. False fix analysis: MAC correction, "worst case" (72 hours of data, Net 2)

Some statistics on the stability of the ambiguity fixing were also carried out, and considered the length (in seconds) of the phase ambiguity fixing, but they are not reported in this paper.

Figure 17 and 18 show the two extreme cases of prediction and detection of FF considering the MAC correction. It is possible to see how, in the "worst case", 86% of the false fixes are identified (about 9 out of 10 FFs), of which 42% are declared as certain. In the "best case" it is possible to see how all the FFs are detected, and 20% of these are declared as certain FFs. It should be emphasized that there is no cause for concern owing to a low rate of false fixes declared as certain: it is important that they are identified as even possible FFs (warning) rather than they are not identified at all. Considering the possibility of reporting the quality of the positioning to the user during the survey, and comparing the quality of the survey using a traffic light sysytem, the epochs when the neural network does not show any abnormality could be identified by a green light, the epochs when there is a warning by a yellow light, and when the FF is declared as certain (red alert) by a red light.

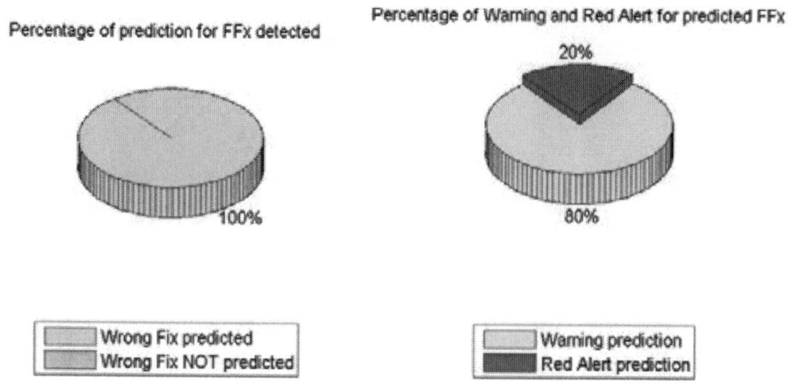

Figure 19. False fix analysis: VRS® correction, "best case" (72 hours of data, Net 3)

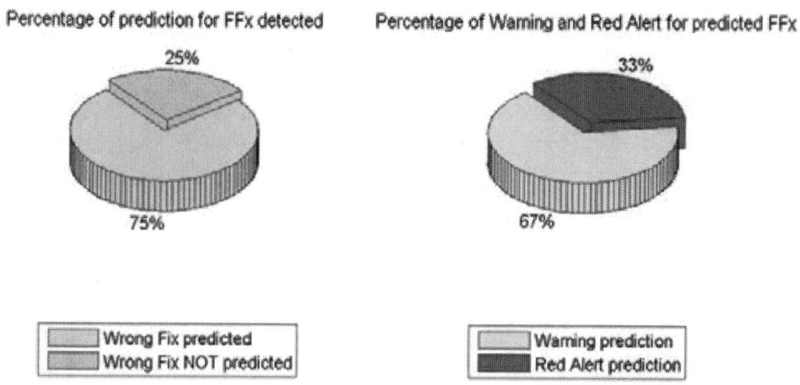

Figure 20. False fix analysis: VRS® correction, "worst case" (72 hours of data, Net 2)

If we consider instead the VRS® correction, it can be seen that in the "worst case", 25% of FFs are not predicted. This percentage is quite worrying as it means that 1 in 4 FFs could not be identified. Of the remaining 75% of predicted FFs, 33% (or rather 1 in 3) are declared as a certain false fix, while 67% are a possible false fix (warning). Another important analysis regards the percentage of "errors" of the neural network, in terms of the number of sure or possible FF that there are not been verified.

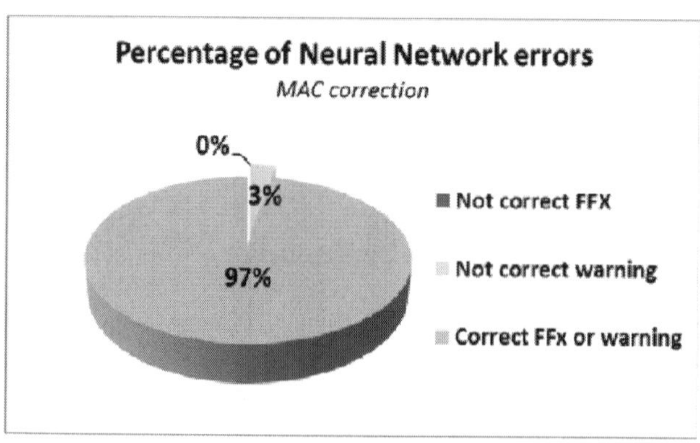

Figure 21. MAC correction: Prediction errors

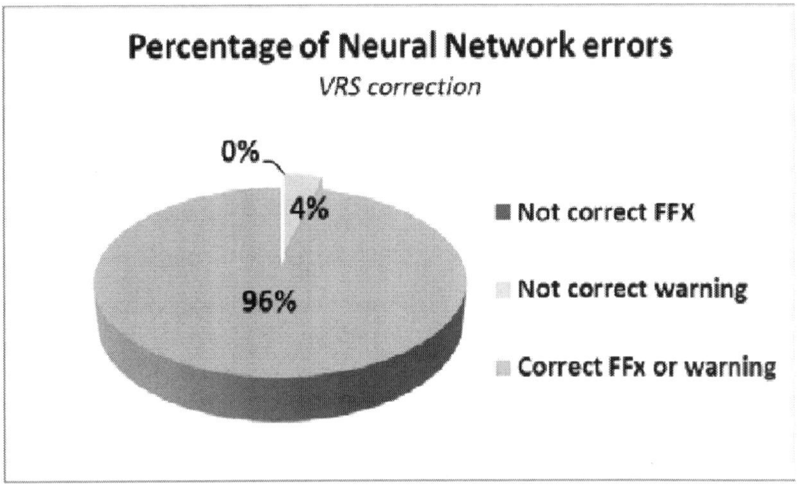

Figure 22. VRS® correction: Prediction errors

As can be seen from Figure 21 and 22, the neural network can be considered efficient, consid- ering the MAC and VRS® corrections respectively, because 3% and 4% of cases report possible FFs that do not really occur, while never sure FF happen. So it is possible to affirm that the type of neural network and its training phase have proved effective. In this phase of the job, we only considered a few parameters obtained from the NMEA message, ignoring all the information derived from the network software that could further improve the results. We, therefore, want to continue in the future, to carry out experiments to try and reduce the 25% of unpredicted FFs that occur, and that cannot be identified by the parameters considered in the current state and, as shown in the graphs, reveal almost certainly a problem in a NRTK network.

7. Conclusion and future developments

As shown in the present work, it is not always possible to identify which parameter (or its variation) leads to a false fix. It is, however, possible to affirm that, according to the studies carried out, the significant quantities that allow the receiver to identify the presence of a FF with high probability, are:

- the latency of correction;
- the noise of the rms value (σ) of the three coordinates;
- the DOP index and therefore the number of satellites used in parallel.

It appears, in the authors' personal opinion, to be of great importance to consider the network parameters (such as ionospheric and tropospheric delays), to try to identify the triggers of FFs— currently unexplained by only considering the parameters — to attempt to reduce quite high rates of unreported false fixings. It should, however, be underlined again that a low rate of FF declared as certain is not a cause for concern. It is important that they are identified even as a possible FF (warning) rather than not be identified.

The input parameters of the network can be expanded. In addition, it is possible to affirm that the network can be trained for a "stop & go" positioning of the rover receiver or, with proper choice of these parameters, for a receiver in motion. The user will choose the type of survey and the receiver and, at the same time, the type of quality control of the survey.

For a quality analysis of the survey to be made in retrospect, it might be important to combine visual information (colour) with the quality of the measurements. For example, it might classify these measures on a screen indicating these points and the respective colour code (from green for the measures considered reliable, up to red for the measures considered unreliable), so the user can later return to measure, perhaps with other conditions of the satellite visibility, the points made previously little or not reliable.

References

1. Dabove, P, De Agostino, M, & Manzino, A M. Achievable positioning accuracies in a network of GNSS reference stations. In: Shuanggen Jin (ed.) Global Navigation Satel- lite Systems: Signal, Theory and Applications. Rijeka: InTech; (2011). Available from http://www.intechopen.com/books/global-navigation-satellite-systems-signal-theory- and-applicationsaccessed 3 February 2012)., 191-217.

2. Dabove, P, & De Agostino, M. What effect does network size have on NRTK positioning?. Inside GNSS (2011). , 24-29.

3. Wübbena, G, Schmitz, M, & Bagge, A. GNSMART Irregularity Readings for Distance Dependent Errors. Geo++ White Paper; (2004).

4. Vollath, U, Buecherl, A, Landau, H, Pagels, C, & Wagner, B. ION GPS (2000). Multi-Base RTK using Virtual Reference Stations: proceedings of the 13th International Technical Meeting of the Satellite Division of The Institute of Navigation, ION GPS 2000, Salt Lake City (UT-USA), September 2000.

5. Wübbena, G, Bagge, A, Seeber, G, Böder, V, & Hankemeier, P. ION GPS (1996). Reducing distance dependent errors for real-time precise DGPS applications by establishing reference station networks: proceedings of the 9th International Technical Meeting of the Satellite Division of The Institute of Navigation, ION GPS 1996, Kan- sas City (KS-USA), September 1996.

6. Fausett, L. Fundamentals of neural networks: architectures, algorithms, and applications. Paris: Lavoisier; (1994).

7. Landau, H, & Euler, H *J. ION GPS 1992:* On-the-Fly Ambiguity Resolution for Precise Differential Positioning: *proceedings of the 5th International Technical Meeting of the Sat- ellite Division of The Institute of Navigation, ION GPS 1992,* Albuquerque, (NM-USA), September (1992).

8. Frean, M. The Upstart Algorithm: A Method for Constructing and Training Feedforward Neural Networks. In: Neural Computation; (1990). , 198-209.

9. http://www.navipedia.net/index.php/GBAS_Systems
10. http://igscb.jpl.nasa.gov/components/prods.html
11. http://www.geopp.de/index.php?bereich=su&kategorie=&artikel=35&seite=2

Section 2
Applications

Chapter-5

L1 GPS as Tool for Monitoring the Dynamic Behavior of Large Bridges

1. Introduction

In 1973 the JPO (Joint Program Office) subjected to United States Air Force received the mission of Department of Defense for implanting, developing, testing and use a spatial positioning system for military applications and able to calculated coordinates and guide missiles according to the project "Star Wars".

The Global Positioning System is fruit of those studies which became able to use the L band of carrier phase (frequency microwaves around 1 up to 3 Ghz and wave length close to 23 cm) for calculating the spatial trilateration. Therefore, in 1978 began the launch of the first NAV- STAR (Navigation Satellite with Time and Ranging) satellites – the begging of GPS as is known today. Due to the high cost of project and as the MIT confirmed by itself the excellence on civil applications (geodesy, topography, navigation, digital modeling, simulators), the American Congress, with the acquiescence of the U.S. President, pressed the Pentagon to open the NAVSTAR system for civil use and other countries. However, only from the 90s is that the GPS became popular. This was a result of technological advance in the micro-computers field allowing the trackers manufacturers to produce the GPS receivers that processed in the receiver, the codes of the received signals.

In this context, from the end of the 80´s, the technology of Global Positioning System, until then, used to conduct surveys of areas, deploy geodetic networks, manage resources, track fleets of vehicles, ships, achieve the control of displacements of structures under static load, etc. began to be used to characterize the dynamic displacement of large structures, earthquakes and so, it was being considered as a tool to extract the values of frequency and amplitude of the displacements with a good accuracy. The amplitude range of displacements detectable with GPS allows to be used as a tool for monitoring the displacement in several kind of structures and the development of sensors of 100 Hz date rate, it will be possible to identify not only the natural frequencies of a structure, but also the frequencies of its several vibration modes.

Goad (1989) conducted the first experiments with the purpose of investigating the feasibility of using GPS to monitor the crest of a dam in Lawrence, Kansas, USA and Lovse et al. (1995) performed the first test to measure the frequency vibration of Calgary Tower in Alberta, Canada, using GPS receivers in differential mode and conventional instruments. The authors verify that the frequency of vibration of the tower of 160 m high, under the wind action was approximately 0.3 Hz.

This time until the present day, several methods were developed and tested the detection of small amplitudes and frequencies of dynamic displacements, but with no results so promising as this that only uses the signal from two satellites through the interferometry technique.

2. Theoretical basis of methods developed

The method applied on this research uses the GPS data supported on the interferometer phase principle. The interferometry is use of the phenomenon of interference between signals, to perform, for example, measures of distance through the phase change caused on one of the both signals. Figure 1 shows, a light beam incident on a mirror is divided into two beams perpendicular to each other. Part of the light beam is reflected and part through another medium. This portion hits a silver mirror and is reflected. Furthermore, the other portion hits another mirror, which can be moved and is also reflected. In this case the beams walk the same distance and the reconstituted light source can be seen reflected on the screen (Holmes, 1998).

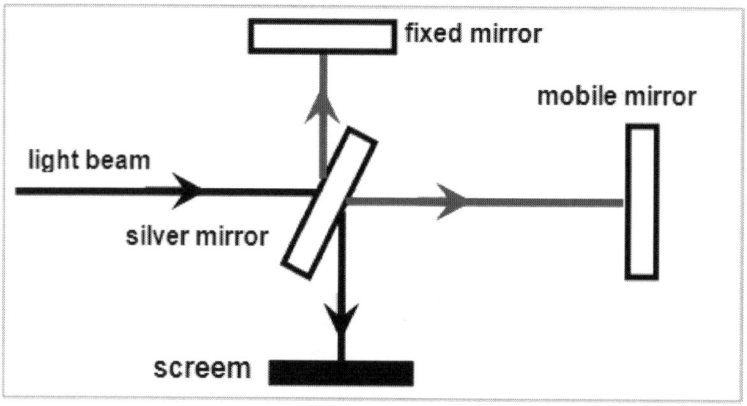

Figure 1. Interferometer scheme (Michelson Interferometer)

If the mobile mirror is moved a distance from its original position, the beam of blue light travel a distance greater than the red beam, causing a different pattern of interference, which can be seen on the screen (Figure 2).

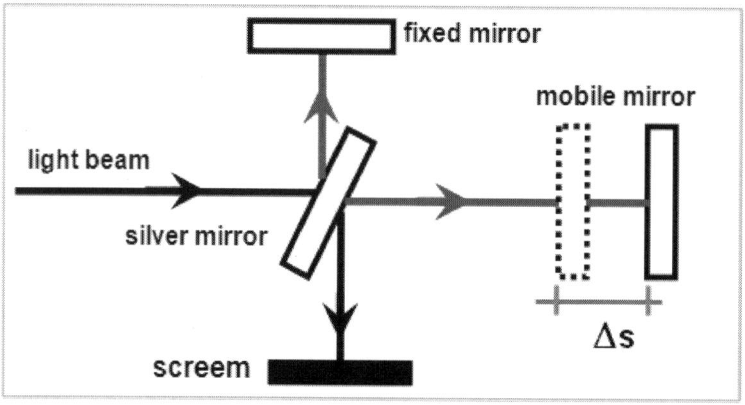

Figure 2. Interferometer scheme with a displacement of one mirror

Considering the beam of light as the electromagnetic wave emitted by the antenna of a GPS satellite and the mirrors with the GPS antenna (Figure 2), vertical or horizontal movements senoidais suffered by the mobile antenna will change the phase of the signal collected by the receiver connected to this antenna. This changes the relationship between phases of the GPS signal received by antenna - mobile and static - since the length of the path is no longer the same. This phase change, then caused by the movements caused by the antenna may have its amplitude and frequency calculated. The mobile antenna, for example, can be fixed in a structure under dynamic loading action.

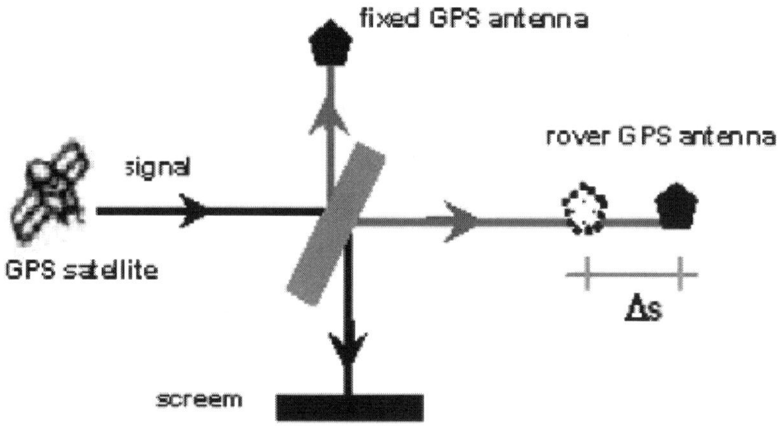

Figure 3. Interferometer of phase related to GPS signal and antennas

Illustrated in Figure 4 is a GPS antenna set in middle of a suspension bridge under dynamic oscillation. The frequency and amplitude of oscillation of the middle can be determined from the analysis of GPS signals collected.

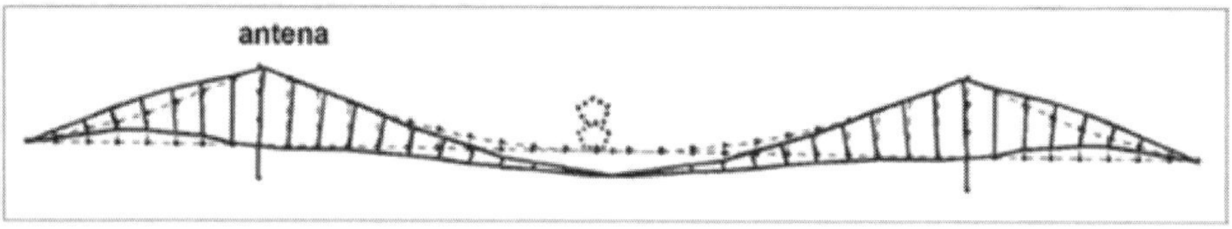

Figure 4. GPS mobile antennas under oscillations of the middle span of a bridge

The method, based on the interferometer phase, requires only the data collection from two satellites, with phase angle of around 90 degrees and not more than a constellation of more than four satellites. Thus, to measure a vertical displacement, for example, is necessary that one satellite be close to 90 degrees and another with elevation next to the horizon (Figure 5). Processing of double difference phase the lowest satellite is considered as the reference satellite, allowing then to obtain the vector of residues of the highest satellite, called here the 'measuring satellite'. With this configuration, there will be a great contribution in the final result of processing the data of double phase difference - residuals - (item 2.1) due to changes in phase, the signal of the satellite close the zenith in relation of the satellite close the horizon, which hardly detects the movement of the antenna.

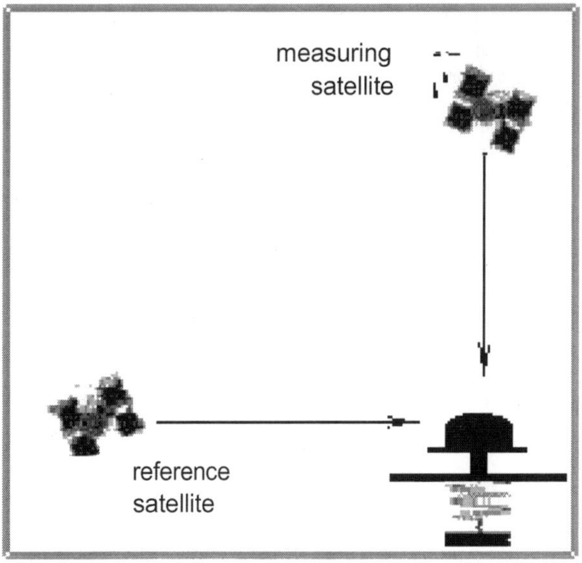

Figure 5. Satellites configuration in relation to a GPS antenna suffering vertical displacements

Similarly, when it is desired to characterize horizontal displacements, the highest satellite is considered as a reference, allowing to obtain the residuals vector of the lowest satellite, will be found where the largest contribution of 'errors' due to changes in signal phase of the satellite close the horizon.

2.1. Double difference phase

The double difference phase, illustrated in Figure 6 is represented by the equation:

$$\nabla\Delta PR^{i,k}_{A,B}(t) = \nabla\Delta \left| R(t) \right|^{i,k}_{A,B} + c\nabla\Delta dT^{i,k}_{A,B} + \lambda \nabla\Delta N^{i,k}_{A,B}(t) + \nabla\Delta\varepsilon^{i,k}_{A,B} \qquad (1)$$

where ∇ an operator which represents the difference between satellite i and k and Δ represents the difference between receivers A and B (Wells, 1986).

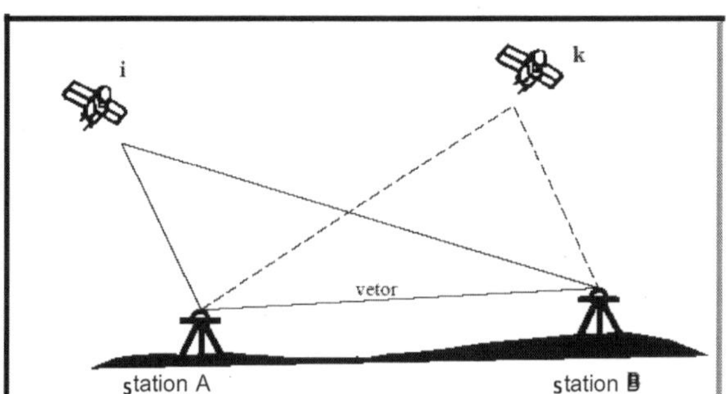

Figure 6. Double difference phase representation

When the baseline done (distance between two points observed with CPS receivers) is short (less than 10 km distance), the errors due to the satellites orbit and errors due to tropospheric and delay ionosferic delay can be removed or reduced almost totally due to similarity of the conditions observed in two points. Furthermore, errors due to multipath, errors due to variation of antenna phase center and (Tranquila, 1986), loss of cycle and other random errors are not removed. And it is possible to access the corresponding value to these errors by the remaining residues of the double phase difference obtained from the adjustment of observa tions by least squares. This method consists on accepted as the best estimate of the redundant observations the value that become minimum the sum of the squares of the residuals. Rewrit ing the eq. [01] in matrix shape it has been:

$$\tilde{N}DF = D \times F \tag{2}$$

where:

D: corresponds to the matrix operator of the double phase difference. The size of the matrix is given by $(R-1)\cdot(S-1)$, $R\cdot S$, where *R* is the number of receivers and *S* is the satellites number.

It is eq. [14]:

$$D\cdot\overset{o}{V}+D\cdot V=D\cdot A\cdot\delta_x \tag{3}$$

or

$$\overset{o}{V}{}'+V'=A'\cdot\delta_x \tag{4}$$

where:

$\overset{o}{V}{}'=D\cdot\overset{o}{V}$: vector of double difference phase

$V'=D\cdot V$: residuals vector of double difference phase

$A'=D\cdot A$: shape of double difference phase matrix

The residuals vector can be written according to the vector of baseline-adjusted subtracting the baseline vector of raw baseline values (without adjustment):

$$V'=L_a-L_b \tag{5}$$

where:

V : residuals vector, ie, difference between adjusted values and the raw values;

$L\,a$: vector of adjusted baselines components;

$L\,b$: vector of baselines components of processed from GPS observations.

2.2. Electro-mechanic oscillator for calibrating vibrations

To calibrate the measurement of dynamic displacements previously unknown was developed an electro-mechanical oscillator which applies controlled movements - on related to displace- ment and the speed of it - on the GPS antenna that will suffer the movements of the footbridge span and the bridge. The oscillator is powered by battery. Figures 7 and 8 present a GPS antenna mounted on the oscillator and a detail of the system that controls the amplitude of the displacement.

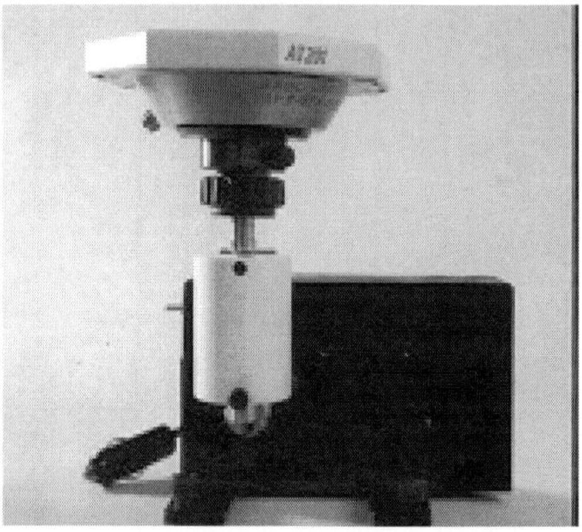

Figure 7. GPS antenna over the electro-mechanical oscillator

Figure 8. Detail of electro-mechanical oscillator

2.3. Spectral Analysis of GPS data

The Fast Fourier Transform was the tool chosen to perform the analysis of the double difference phase residuals, here called raw residuals, in the frequencies domain and consequently, to identify the corresponding frequencies due to periodic displacements in a spectrum that also presents the very low frequencies due to multipath of the environments highly noisily and others - effects of variation of the antennas phase center - which is accentuated in highly reflective environments and in a not static observations (Wells et al., 1986).

The essence of the Fourier Transform (FT) of a wave is to decompose or separate it into a sum of different frequencies senoides. If the sum of these senoides results in the form of original wave, then was given its Fourier Transform. A function of wavelength, in the time domain is transformed to the domain of frequencies, where is possible to determine the magnitude, frequency of the wave and perform the filtering of undesired frequencies (noise), Brighan (1974).

3. Tested structures for the method analysis

The dynamic analysis of a structure aimed to determine the maximum displacements allowed by the project (design constraints), speed and accelerations (comfort for users), internal efforts, stress and deformations (fatigue of the material that composes the structure) (Laier, 2000). Thus, the analysis can diagnose the actual state of structure conservation (regardless of external appearance), predict its life time and determine economic solutions of recovering in order to prolong the durability of the structure.

Two bridges were submitted to dynamic tests – mobile load - to test the GPS as tool for monitoring structures.

The first structure tested was a cable-stayed wood footbridge built in Sao Carlos Engineering, University of São Paulo, São Paulo state, Brazil, in 2002 (Figure 9), Pletz (2003). The footbridge, which is presented as the first wood footbridge built in Brazil with the deck in curve shape, has a 35 meters total length deck on *Pinus taeda* wood and wood used for the tower was Eucalyptus citriodora. The tower consists of a pole with thirteen meters long, 55 cm in diameter at the base and 45 cm at the top. The footbridge was divided into seven modules with nominal dimensions of 5 m in length, 2 m wide, 20 cm in height, each consisting of 37 slides measuring 5 cm in width and 20 cm high and 5 m in length (Pletz, 2003).

Figure 9. Cable- Stayed wood footbridge in Sao Carlos Engineering School, USP

The second structure where the method was tested is the Hawkshaw Bridge (Figure 10), a cable-stayed bridge with cables anchored in a harp shape. The bridge is located in the province of New Brunswick, Canada, at Hawkshaw Bridge Road, 0.20 km North, at the intersection with Highway 2, in the Nackwic district and it link the two shores of the Saint John River.

The Hawkshaw Bridge is composed by a steel deck i-beam, with three spans. Inaugurated in 1967, the bridge has its longitudinal axis predominantly in the north-south direction. The center span has 217 m in length, the north direction span has 29.44 m and the south has 54.44 m, a total of 301.20 m. The deck is supported by two steel towers with 36 m in height where six sets of steel cables are fixed on each side. The board has width of 7.90 m, with two traffic lanes in opposite directions (Figure 10).

Figure 10. Hawkshaw Cable-Stayed Bridge, New Brunswick, Canada

The footbridge and the bridge have in common the constructive system, based on the theory of cable-stayed. A description of the structural function of this type of structure is described below.

3.1. Structural behavior of tested structures

The cable-stayed bridge can be defined as a structure composed by a main beam supported by steel cables tensioned, stuck on top of one or more towers, having then two or more spans. A good comparison for understanding the structural system of a cable-stayed bridges is to imagine that the arms of the human body is the board of the bridge and the head becomes the tower, creating two identical spans in length (the arms) and the muscles support the arms (Figure 11). With a piece of rope, of 1m length, tie the two elbows and puts the middle of the rope on the head. Thus, the string will act as a cable stayed that supports the elbow. With a second piece of rope, of 1.5 m long cable is then the two pulses. There was the same way, placing the rope on the head, there is another cable stayed. The strength or compression of the two cables that support the arms (board) is felt in the human body on the head, then, the bridge tower. Thus, as the cables can be used as intermediate supports for beams, the concept of bridge estaiada can overcome long spans. The cables are then the most important elements of a bridge estaiada because they, under stress, support the weight of the beams and transfer efforts to the mooring system, the fixed towers, suffering compression (Tang et al. (1999); http:// www.pbs.org/wgbh/nova/bridge/meetcable.html).

Figure 11. Concept of cable-stayed bridge

4. Tests on a cable-stayed wood footbridge

According to the objective of to monitor the dynamic behavior of the footbridge were planned forced vibration tests to be made with pedestrians walking over the deck. Each test lasted approximately 10 minutes, enough time to excite the several vibration modes of the footbridge.

4.1. Instrumentation

The instrumentation consisted of a pair of GPS receivers with 20 Hz data rate with choke ring antennas and a transducer of displacement Kiowa DT 100, with Vishay data acquisition system of 20 channels and 10 Hz data rate, model 5100B Scanner. Figure 12 illustrates the layout of the instruments used on a footbridge and Figures 13 and 14 illustrate the layout of these instruments in the footbridge. The electro-mechanical oscillator was adjusted to apply a displacement with amplitude of 12 mm and frequency of 1.0 Hz.

Preliminary tests with pedestrians walking on the bridge were made for a rough knowledge of the dynamic behavior of the bridge, monitoring the extent and frequency of vertical displacement, which was conducted by two observers. One observer, using a total station and a piece of tape measure set in the center of the leg 2 determined the average amplitude of the approximate vertical displacement caused by pedestrians. Another observer determined with the aid of a stopwatch, the approximate frequency of the footbridge at the same time. And the

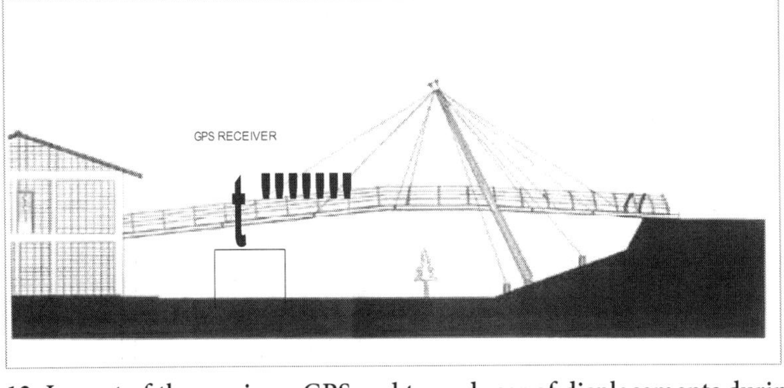

Figure 12. Layout of the receivers GPS and transducer of displacements during trials

Figure 13. The GPS receiver and transducer of displacement installed in a footbridge

values obtained for the scale ranged between 8 and 12 mm and frequency varied in the range from 100 to 120 cycles per second, this value of frequency of vibration induced by pedestrians, as set out by Pretlove et al. (in CEB, 1991).

During the tests, the antenna was set in electro-mechanical oscillator to have a peak in the spectrum of known frequency and amplitude, and then serve as a calibrator for the peak due to the displacement of the bridge. Thus, the oscillator was adjusted to apply movement amplitude of 3 mm, with a frequency of 1 Hz at the antenna. The forced vibration tests were performed using a mobile people and cargo, which walked in an orderly way on the board (Figures 15 and 16).

Figure 14. Transducer of displacements and data acquisition system

4.2. Footbridge tests results

The data processing collected with GPS and transducer displacement are described below.

Figure 17 presents the data obtained with the transducer displacement during the filed test carried out in a footbridge with pedestrians on moving - figures above -. By means of maximum and minimum values recorded it was determined the amplitude of displacement, resulting in 13 mm.

Applying the Fast Fourier Transform (FFT) to these values in the spectrum, it can clearly see the peak corresponding to the periodic movements recorded by the displacement transducer, with frequency of 2.0 Hz (Figure 18).

During the tests carried out, the measuring satellite and the reference were close to the 74 degrees (PRN 28) and 05 degrees PRN (26), respectively. Figure 19 illustrates the residuals of double phase difference between these satellites. The spectral analysis of these residues (Figure 20) allows extracting the frequency value of the electro-mechanical device, 1.1 Hz and value of the footbridge, 2.1 Hz, under forced vibration. In Figure 20 it can be observed also that the corresponding peak to the displacement applied by the oscillator at the antenna is perfect

Figure 15. Front view of pedestrians walking on a footbridge and view of GPS antenna over oscillator

Figure 16. Side view of pedestrians walking over a footbridge

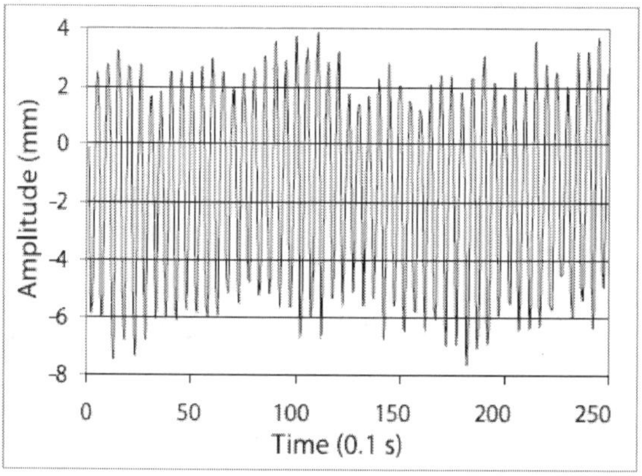

Figure 17. Data collected by transducer of displacement

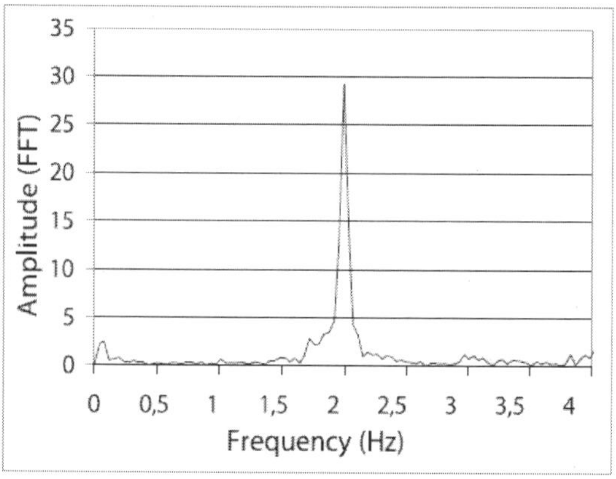

Figure 18. Power spectrum of data collected by transducer of displacement

because the movements are applied by a machine - the electro-mechanical device. Already, the peak corresponding to the displacement caused by the pedestrian's action of walking, although ordered, is not perfect, because each person has a step length and a certain weight.

Through the direct comparison of the peaks of amplitudes illustrated in Figure 20, it was determined the value of the displacement amplitude value of the footbridge, because it is known that the corresponding value to the amplitude displacement applied by the oscillator was 12 mm. Measuring up in Figure 20, it was obtained:

$$\left.\begin{array}{l} 8.3\text{ cm} \;\text{¾¾}\; 12\text{ mm} \\ 9.0\text{cm} \;\text{¾¾}\; x \end{array}\right\rangle x \quad 13\text{ mm} \tag{6}$$

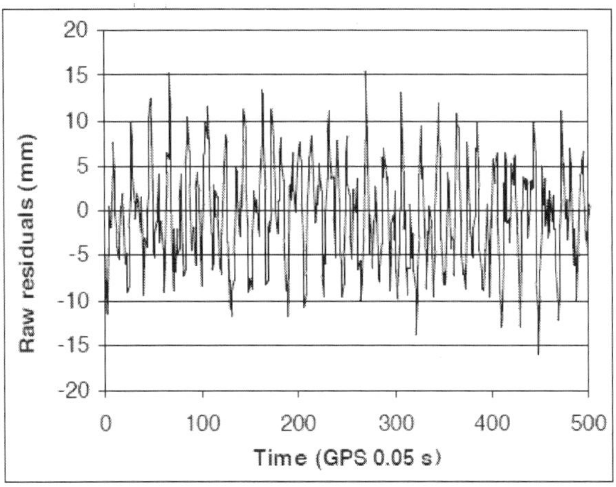

Figure 19. Raw phase residuals of the dynamic vertical displacement response

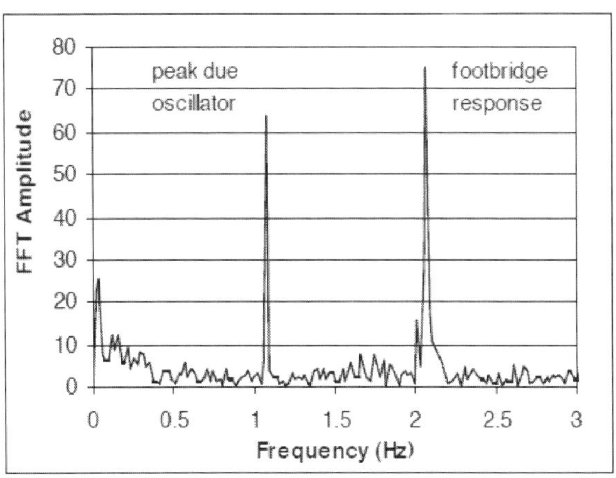

Figure 20. Power spectrum of the raw residuals with peaks due to vertical dynamic displacements applied on the GPS antenna

The comparison of the values obtained with GPS and with the transducer can be summarized in the table below.

Two other tests were performed with 12 minutes duration, with the objective of trying to detect the natural frequency of the footbridge and the harmonic frequencies of other vibration modes.

During these two tests, the reference satellite, PRN 29, was at 16 degrees of elevation and measuring satellite, PRN 28, was at 78 degrees and the footbridge was not instrumented with displacement transducer. The spectral analysis showed the occurrence of more two peaks, besides the peak due to oscillations caused by the electro-mechanical device in the antenna, with a frequency value of 1.1 Hz and the peak due to the action of walking with a frequency of 2.0 Hz, illustrated in Figure 20. In Figures 21 and 22, below, illustrates besides these two peaks, the peak of frequency value equal to 3.1 Hz, corresponding to the value of the natural vertical frequency according to Pletz (2003), which presents to the same frequency, the value of 3.2 Hz obtained by the Finite Elements analysis and the fourth peak, with a frequency value equal to 4.1 Hz, corresponds to the frequency of the first vibration mode.

Table 1. Values of natural frequency and amplitude of displacement obtained with the GPS and the transducer

Equipment	Frequency of footbridge response to pedestrians walked (Hz)	Vertical amplitude displacement (mm)
GPS – 20 Hz e antena Choke ring	2.1	13.0
Transducer displacement -100 mm	2.0	13.0

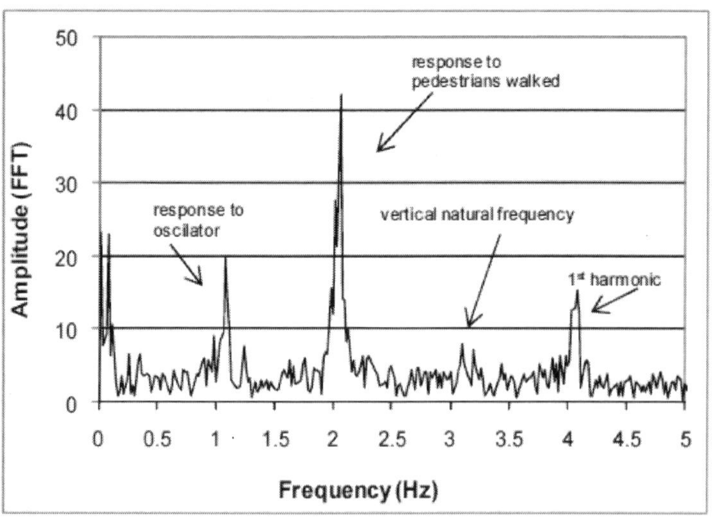

Figure 21. Frequency detected by GPS data – first test

5. Tests on a Hawkshaw cable-stayed bridge, Canada

With the objective to monitor the dynamic behavior of a Hawkshaw bridge it was planned forced vibration tests carried out with trucks – design trucks - on the deck during the author's doctoral internship at the University of New Brunswick, Canada in October 2003. The trial was supported by the Department of Geodesy and Geomatics Engineering of University of New Brunswick, Geodetic Research Laboratory, Canadian Center for Geodetic Engineering and the New Brunswick Department of Transportation (NBDT).

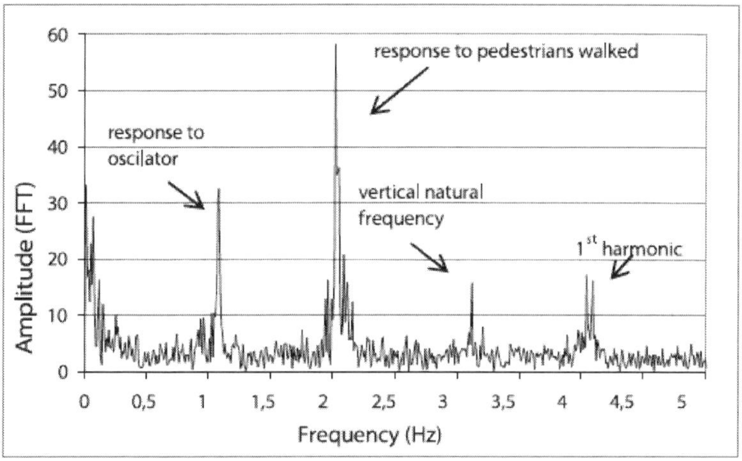

Figure 22. Frequency detected by GPS data - second test

5.1. Instrumentation

The Figures 23 and 24 below illustrate the layout of equipments installed. Two GPS receivers that were the reference stations were installed on top of a gravel mountain, 30 m from the end of the bridge going to south span, which is the highest place close to the bridge (Figures 25 and 26). For each fixed station, REF 2 and REF 3, it was used a Novatel OEM4-DL4 receiver with Pinwheel antenna and a Trimble 5700 receiver with Geodetic

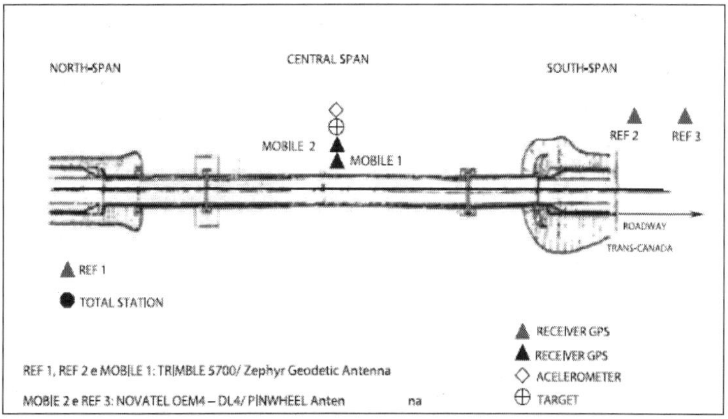

Figure 23. Layout of instruments used on the monitoring of a bridge

antenna ZephyrTM. In the bodyguard of the central span were installed two GPS receivers on the electro-mechanical device to register well known oscillations besides of the bridge (Figures 27and 28). All receivers were programmed to collect data with a 5Hz rate. A total station was used to perform measurements with the design-trucks on a deck and also an accelerometer for measurement of the frequencies vibration of the bridge's deck.

Figure 24. General view of a restricted railway for installing equipments on a central span. On left, Ronald H. Joyce (Senior Technical Advisor – Maintenance and Traffic) and on right Neil Hill (Bridge Superintendent)

Figure 25. Reference stations used for monitoring the bridge

Figure 26. Detail of reference stations used for monitoring the bridge

Figure 27. Wood support for electro-mechanical device

Figure 28. View of accelerometer fixed on wood support

Figures 29 and 30 illustrate some examples of-trains that were used as mobile load during the dynamic test on the bridge deck.

Figure 29. Design truck

5.2. Results of tests in a Hawkshaw Bridge

Figure 30. Design truck

5.2.1. Monitoring of vertical displacement of the deck

During one of several tests performed on the bridge, the measuring and the reference satellites were close to 81 degrees of elevation (PRN 02) and 09 degrees (PRN 31), respectively. Figure 31 illustrates the residuals of double difference phase of all satellites in relation to PRN 31, since it was looking for at that time, to check the vertical dynamic behavior of the central span when it allowed that four design-trucks crossed the bridge. The crossed takes nearly 75 seconds. In Figure 32 it is possible to see clearly the graphic description of vertical displacement of the instrumented middle span section that reached 8 cm amplitude. Therefore, other four design-trucks were asked to stop in the middle of the central span to take the measures with the Total Station, obtaining a mean value 8.3 cm.

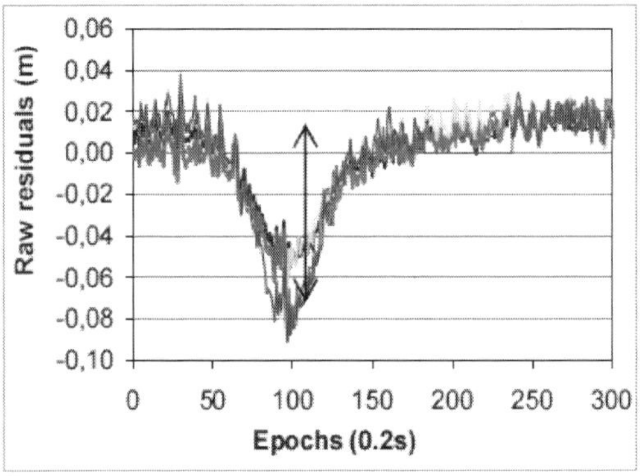

Figure 31. Raw residuals of vertical displacement of central span – 4 design-trucks

5.2.2. Monitoring of lateral displacement of the deck

During a second test, the reference and measuring satellites were close to the 80 degrees of elevation (PRN 02) and 06 degrees of elevation (PRN 31), respectively. Figure 33 illustrates the residuals of double difference phase for all satellites in relation to the PRN 02, as it was looking for by a lateral dynamic behavior of a central span. And in Figure 33 is possible to see clearly

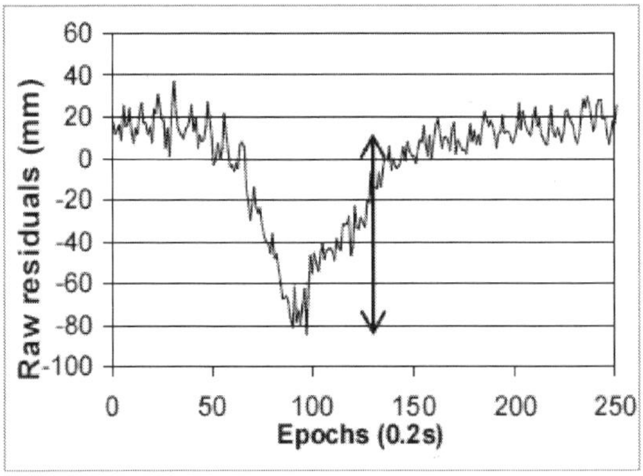

Figure 32. Raw residuals describing the vertical displacement of central span

the graphic description of the lateral displacement of the instrumented section in the central span, where a design-truck of 60 tons crossed the bridge. The crossed takes nearly 45 s.

Figure 34 illustrates only the residuals of the lowest satellite (PRN 31) for better visualization of the lateral dynamic displacement caused by a mobile load of 60 tons. The spectral analysis of these residuals (Figure 35) allows extracting value of the deck´s lateral frequency vibration when subjected to vibration caused by a mobile load. The lateral frequency value of the deck was 0.60 Hz. The amplitude of dynamic displacement showed the average value of 3.5 cm. Furthermore, the lateral displacement of the board, when the truck starts to cross, reaches the middle of the deck and starts to exit the bridge, has average amplitude of 3.5 cm.

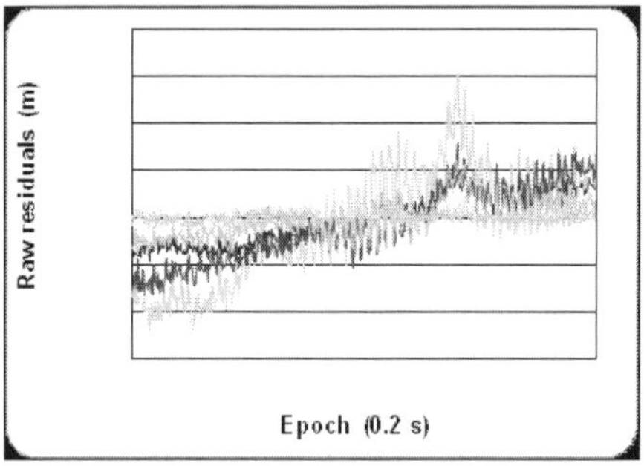

Figure 33. Raw residuals of lateral dynamic displacements of deck

Spectral analysis of these residuals by FFT allows extracting the lateral frequency vibration value when subjected to vibration caused by a design-truck of 60 tons crossing the deck. The lateral frequency of the deck was 0.60 Hz (Figure 35).

5.3. Team which collaborated for performing the tests

Figures presented below are the people who helped to perform the test. The traffic controllers from NBTD in Figure 36 and in Figure 37 can be seen by the technicians who collaborated for performing the tests - Howard Biggar (Geomatics Technologist – GGE – UNB), E. Daniel Wheaton (Chief Technician Civil Engineering- UNB) e Jason D. Bond (PhD Candidate - CCGE - UNB).

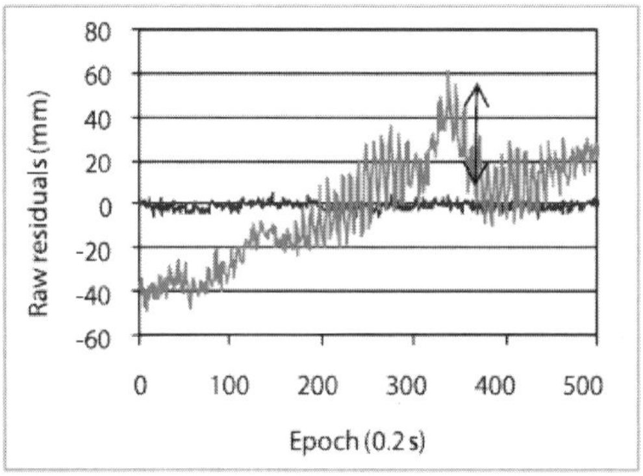

Figure 34. Raw residuals describing the lateral dynamic displacements of deck

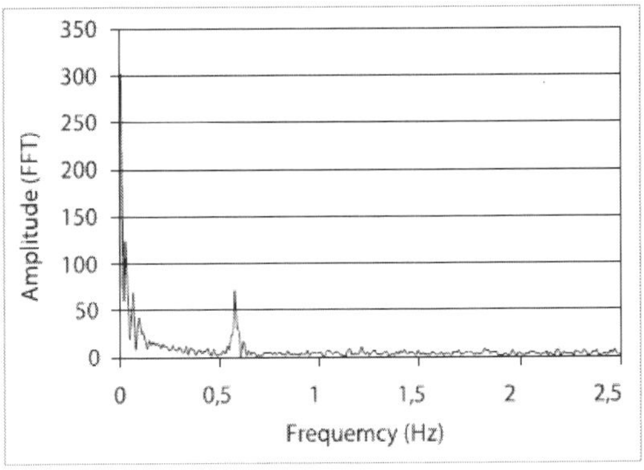

Figure 35. Spectrum of residuals with the peak due to lateral frequency vibration of central deck

Figure 36. From left to right: Howard, Daniel, Ana Paula, Neill and traffic controlers from NBDT

Figure 37. From left to right: Daniel, Howard, Jason and Ana Paula (author)

6. Conclusions

Based on studies, field experiments and analysis of results presented in this research, it can be concluded that the GPS data collection researched method has been established because its capability and efficiency. And therefore it ensured to GPS the label of monitoring instruments and characterization of the dynamic behavior of structures.

A comparison of results obtained with the GPS and the transducer displacement, resulting from vibration tests conducted on a wood cable-stayed footbridge to confirm the reliability of the results obtained by GPS to characterize the dynamic behavior of structures, which agreed satisfactorily with the values by the Finite Elements theory and theoretical values of CEB (Comite Euro-International du Beton - Bulletin D′ Information no. 209). Therefore, the results proved the efficiency and capability of the collection method and GPS data analysis to obtain the frequency values and amplitude of dynamic displacement, showing that the limitation imposed by the necessity of a particular satellite geometric configuration, in this case did not prejudiced the program and performing the tests. As the method does not required a good geometric distribution of satellites - and only two satellites -, allows obtaining reliable results on the dynamic behavior of a structure in any latitude of the globe. The use of electro- mechanical oscillator as a calibrator was reliable, providing also to produce calibrated values of frequencies and amplitudes of unknown displacements, since the electro-mechanical oscillator can be used to produce known oscillations.

The results of the second test of on using the method of this research on a large man-made road structure, the Hawkshaw Cable-stayed Bridge showed the full possibility on using GPS for characterizing the dynamic behavior of this type of structure.

Given the above, it was concluded that GPS, or the method of data collection employed, allows for the graphical description of the dynamic displacement amplitude of the middle span deck and the identification of modal frequencies of bridges under the controlled traffic action or not, may be used by engineering as a tool for monitoring structures.

References

1. Brigham, E. O. (1974). *The Fast Fourier Transform*. 1. ed. New Jersey. Prentice-Hall, Inc.. Cap. 1, , 1-8.

2. Cosser, E, Roberts, G. W, Xiaoling, M, & Dodson, A. H. (2003). *The Comparison of Single Frequency and Dual Frequency GPS for Bridge Deflection and Vibration Monitoring*. Proceeding, 11 FIG Symposium on Deformation Measurements, Santorini, Greece.

3. Holmes, J. B. (1998). http://www.cbu.edu/~jholmes/252Light21n/sld040.htm.

4. Ko, J. M, Ni, Y. Q, Wang, J. Y, Sun, Z. G, & Zhou, X. T. (2000). *Studies of vibration-based damage detection of three cable-supported bridges in Hong Kong*. Proceedings of the Inter- national Conference on Engineering and Technological Sciences Session 5: Civil Engi- neering in the 21st Century, J. Song and G. Zhou (eds.), Science Press, Beijing, China, 105-112., 2000.

5. Langley, R. B. (1997). GPS Receiver System Noise. GPS WORLD- Innovation. June 1997.

6. Larocca, A. P. C. (2004a). *O Uso do GPS como Instrumento de Controle de Deslocamentos Dinâmicos de Obras Civis- Aplicação na Área de Transportes.* São Carlos, 2004. 203p. Tese (Doutorado)- Escola de Engenharia de São Carlos, Universidade de São Paulo.

7. Larocca, A. P. C, & Schaal, R. E. (2002). Degradation in the Detection of Millimetric Dynamic Movements due to Metallic Objects close to the GPS Antenna. In: COBRAC 2002, Florianópolis, 2002. CD-ROM. Santa Catarina- Brasil, Artigo n. 20.

8. Larocca, A. P. C. (2004b). *Using High-rate GPS Data to Monitor the Dynamic Behavior of a Cable-stayed Bridge.* ION GNSS The 17 International Technical Meeting of the Satel- lite Division of Navigation. September 21-24, 2004,Long Beach, CA, USA., 2004.

9. Leick, A. (2004). GPS satellite Surveying. New York, John Willey, 2004.

10. Ogaja, C, Rizos, C, & Wang, J. (2001). *A Dynamic GPS System for On-Line Structural Monitoring.* Int. on Kinematic System in Geodesy, Geomatics & Navigation (KIS 2001), Banff, Canada, June. (Download PDF)., 5-8.

11. Pletz, E. (2003). Cable-stayed footbridge with stress laminated timber deck composed of curved modules. São Carlos, 2003, Tese de Doutorado, Escola de Engenharia de São Carlos, Universidade de São Paulo.

12. Pletz, E. (2003). Passarela Estaiada com Tabuleiro de Madeira Laminada Protendida em Módulos Curvos. São Carlos, 2003. 164p. Tese (Doutorado)- Escola de Engenharia de São Carlos- Universidade de São Paulo.

13. Podolny, J. R. W, & Scalzi, J. B. (1976). *Construction and Design of Cable-Stayed Bridges.* New York: JOHN Wiley & Sons. Cap.1, Cap.12, p. 425-453., 1-23.

14. Pretlove, A. J, & Rainer, H. (1991). *Vibrations Induced By People.* Bulletin d' Information In: CEB, 1991. Lausanne. Cap. 1, (209), 1-10.

15. Pretlove et al(1991). V*ibration Problems in Structures. Bulletin* d' Information In: CEB, 1991. Lausanne. Appendix G, (209), 199-202.

16. Pretlove, A. J, et al. (1991). *Bulletin* d' Information Problems in Structures. In: CEB *Vibrations Induced By People.* Lausanne. Cap. 1,(209-Vibration), 1-10.

17. Radovanovic, R. S, & Teskey, W. F. (2001). Dynamic Monitoring of Deforming Structures: GPS versus Robotic Tacheometry Systems. The 10th International Symposium on Deformation Measurements (FIG), Orange, California, USA, March., 19-22.

18. Roberts, G. W, Meng, X, & Dodson, A. H. (2001). *Data Processing and Multipath Mitigation for GPS/Accelerometer Based Hybrid Structural Deflection Monitoring system.* De- formation Measurements and Analysis. ION GPS The 14th International Technical Meeting of the Satellite Division of the Institute of Navigation, 11-14 September 2001, Salt Lake City, USA., 2001.

19. Roberts, G. W, Meng, X, & Dodson, A. H. (2002). *Using Adaptive Filtering to Detect Multipath and Cycle Slips in GPS/Accelerometer Bridge Deflection Monitoring Data.* FIG XXII Internacional Congress, Washington, D.C. USA, April., 19-26.

20. Schaal, R. E, & Larocca, A. P. C. (2001). *A Methodology to Use the GPS for Monitoring Vertical Dynamic Subcentimeter Displacement.* Digital Earth Fredericton, New Brunswick- Canadá- 28, Junho., 2001.

21. Schaal, R. E, & Larocca, A. P. C. (2001). *A Methodology to Use the GPS for Monitoring Vertical Dynamic Subcentimeter Displacement.* Digital Earth Fredericton, New Brunswick- Canadá- 28, Junho., 2001.

22. Schaal, R. E, & Larocca, A. P. C. (2002). A Methodology for Monitoring Vertical dynamic Sub-Centimeter Displacements with GPS. *GPS Solutions*, n.3, Winter., 5, 15-18.

23. Schaal, R. E, & Larocca, A. P. C. (2002). A Methodology for Monitoring Vertical dynamic Sub-Centimeter Displacements with GPS. GPS Solutions, n.3, Winter., 5, 15-18.

24. Shun-Ichi NakamuraP. E. ((2000). GPS Measurement of Wind-Induced Suspension Bridge Girder Displacements. *Journal of Structural Engineering,*, 126

25. Sumitro, S. (2001). *Long Span Bridge Health Monitoring System in Japan.* Presented in a Workshop Sponsored by the National Science Foundation on Health Monitoring of Long Span Bridges, University of California, Irvine Campus, March, 2001. http://www.krcnet.co.jp/papers/pdf/International/UCI2001_sumitoro.PDF, 9-10.

26. Tang, M. (1999). Suspension Bridges. In: Wai-Fah, C. et al. *Bridge Engineering Handbook.* CRC Press LLC. Cap. 19, , 19-11.

27. Tranquilla, J. M. (1986). Mutlipath and Imaging Problems in GPS Receiver Antennas. Presented at Symposium on Antenna Technology and Applied Electromag VI.3.

28. Troitsky, M. S. (1977). Cable-stayed Bridges, Theory and Design. Wiliam Clowes & Sons, Limited. Cap. 1, , 1-10.

29. Tsakiri, M, Lekidis, V, Stewart, M, & Karabelas, J. (2003). *Testing Procedures for the monitoring of Seismic Induced Vibrations on a Cable-Stayed Highway Bridge.* Proceeding, 11th FIG Symposium on Deformation Measurements, Santorini, Greece, 2003.

Chapter-6

Intelligent Traffic System: Implementing Road Networks with Time-Weighted Graphs

1. Introduction

This section introduces the chapter's main subject. First, spatial temporal and spatio-temporal databases are presented as the main databases used in all Geographical Information Systems (GISs) including driving direction systems. Second, a brief introduction of GIS is presented. Finally a brief overview of existing driving path application is included as applications similar of the work proposed here.

1.1. Spatial databases

As most kinds of applications need databases, Geographical Information Systems use what is called spatial databases. Spatial (from space) databases are databases used to store information about geography like: geometries, positions, and coordinates. Also they include spatial operations to be applied on such kinds of data like distance, area, perimeter, direction, and overlap of geometries.

Who might use spatial databases? The following do use spatial databases:

- Mobile phone users: where is the nearest gas station?
- Army field commander: enemy movement?
- Insurance risk manager: houses to be affected?
- Medical doctor: same treatments in some area?
- Transport specialist
- Sports: what seats have good view?
- Emergency services: locate calls.

1.2. Temporal and spatio-temporal databases

A database contains information and transactions that need to be stored, manipulated and retrieved. A number of computer applications that use databases deal only with the most recent or current data. Such a database is called snapshot databases, which represents the data at the current time with recent values. On the other hand, in some applications, users might need to know not just the current information but also past or future information as well. Databases dealing with past, present, and future data are called temporal databases or historical data- bases [5]. Such databases support applications such as managerial information, and geograph- ical information. An example of this is a bank account where the customer might need to know old balances or old transactions. In such a case, users need a history of the information. These databases process a temporal dimension to store and manipulate time-varying data. In temporal databases, time is involved in two common ways [5]:

- Time stamp: one attribute is used to save the time of the validity
- Time interval: two attributes are used to indicate the interval of validity (start time and end time).

It is obvious that most spatial databases are also changing with time. For example, road networks will be constantly changing, areas of lands on maps may change, a moving plane will change its position over time, and so on. In such a case, we have spatial or geographical data that is changing with time. In this case spatial database becomes temporal as well and is called spatio-temporal database. Such databases are categorized in three main categories:

- Discretely changing spatio-temporal databases: like the changes of the geometry of land parcels where it could occur once every year or more.
- Continuously changing spatio-temporal databases: mostly deal with moving objects like plane or cars where the position of the object is changing continuously with time.
- Third category that is a combination of the above.

In reality, most spatial databases change with time and hence most of them are spatio-temporal databases.

1.3. Geographical information systems

Geographic Information System (GIS) is a collection of computer hardware, software for capturing, managing, analyzing, and displaying all forms of geographical information [5].

Geographical Information Systems are being involved in most aspects of life and businesses. All GIS's use spatial databases as their data warehouse that are manipulated and presented in a user interface. Later in this chapter, driving direction example queries are given as examples of GIS applications.

1.4. Driving path and GNSS as a GIS applications

Global Navigation Satellite System or GNSS is a satellite navigation system that uses satellites to provide autonomous geo-spatial positioning with global coverage. It allows small devices (especially mobile) to receive and determine their location (longitude, latitude, and altitude) to within few meters using time signals transmitted along a line-of-sight by radio waves from satellites. These devices calculate the precise time as well as position, which can be used as a reference for scientific experiments.

Finding the driving path is one of the most frequent queries in GIS applications. There are many factors that influence the criteria of finding the driving path; the following are the most important:

- Distance: What is the distance between the origin and destination?

- Road situation: is the road closed?
- Road traffic: how much traffic is on the road?

The chapter is organized as follows: this section introduces the subject in general. Section 2 presents some related and previous work including widely used applications. Section 3 presents our intelligent traffic system as the main solution of the problem at hand and finally section 4 discusses some conclusions.

2. Related and previous work

An overview of related practical and theoretical related work is presented in this section. Example queries are also illustrated. Moreover, the section briefly presents existing artificial intelligence in such applications. Finally, A*Traffic is presented as the main algorithm used in this chapter where a proposal of a time-weighted graph is used as the main data structure where A*Traffic is applied.

2.1. Driving direction applications: Google earth as an example

This sub-section presents one of the most widely used applications for finding driving directions: Google Earth. As a note, the application is not only used as driving directions application only but also offers other GIS services which are out of the scope of this chapter.

2.1.1. Google earth

Google Earth [9, 10] is a geographical system that offers satellite images of the locations along with spatial information such as coordinates and elevation. It contains about 70 TB of Data [10]. It provides three main kinds of data: Raster data, Spatial and Non-spatial data, and Video. Moreover, Google Earth offers a set of functionalities, an important subset of which is:

- Answers location queries: the user gives a location (New York, USA) as input and gets a geographical image as an output. The image can be explored in details; this feature includes: cities, businesses, public places, etc.
- Shows directions: the user gives an origin and destination as input and gets a map as output showing the directions with driving hints on the map.
- Displays spatial information: Google earth shows spatial information like coordinates, elevations, altitude, and others.
- Has learning abilities: Google Earth saves recently and regularly visited locations and queries so that the user avoids delays the next time these locations or queries asked.
- Includes pre-known locations: Google Earth offers a list of the most used locations like government offices, schools, etc.
- Provides user interaction: the user is able to put place marks on the map so he/she can avoid delays the next time visits the same place is visited.
- Provides prepaid online service that provides the customer with live video (with restriction and delay due to security) of any place in the world.
- Other products like Google Earth PRO: it is a paid service that makes it very easy to research locations and present discoveries. In just a few clicks, the user can import site

plans, property lists or client sites and share the view with his/her client or colleague. Moreover, the user can export high-quality images to documents or the web.

In addition to introducing Google Earth, this section present a driving directions query as an example of the driving direction services that Google Earth offers, which is directly related to the work in this chapter.

2.1.2. Driving direction query by google earth: Form "New York, NY" to "Jersey City, NJ"

Figure 1 shows the driving directions with a map image from New York, NY to Jersey City, NJ. The figure shows the input (origin and destination) and outputs on the map (Roads and driving directions). In such queries, "Google Earth" provides driving tips to be followed when driving from the origin to the destination given in the query. The user can be more specific by passing a full address (building, street, city, state, and zip assuming U.S.A. is the country.)

2.2. Artificial intelligence and driving directions

Artificial inelegance is used in graph searching algorithms. Russel and Norving [4] presents several intelligent graph searching algorithms. Here are two important ones:

❖ Greedy best-first search

❖ A* search

The main idea behind these algorithms is that they do not try all possible cases to give an answer. The algorithms use heuristic function to un-check some of the paths. This saves huge amount of time but does not guarantee a best path. However, finding a good heuristic function could guarantee up to 95% finding the best path. This section includes the A* search algorithm to be used in the solution approach presented in this chapter.

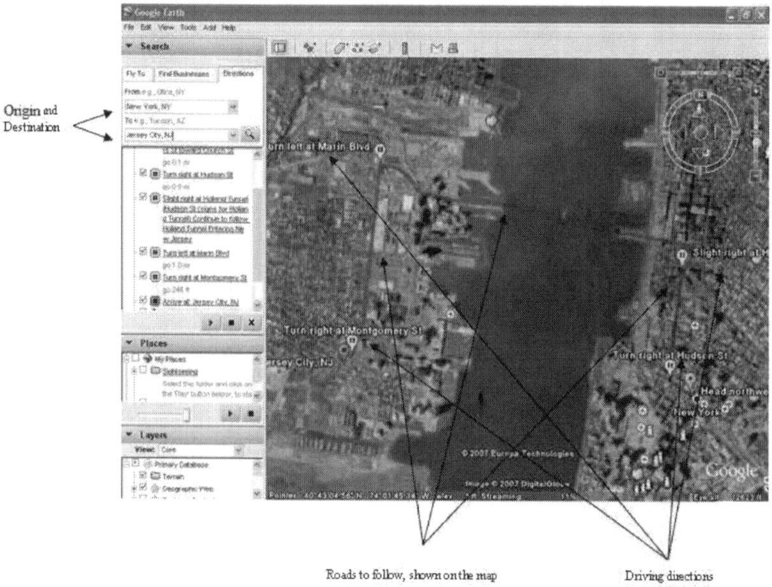

Figure 1. Road and driving directions between New York, NY and Jersey City, NJ

2.3. A* traffic: Design, algorithms and implementation for a driving direction system

This section presents the application algorithms and the application of the intelligent driving path application used in our previous work, which is extended in this chapter. Examples of executions demonstrated using our testing tool, are presented.

2.3.1. A*: an artificial inelegant algorithm for finding driving directions

A* [2, 4] is an Artificial Intelligent graph algorithm proposed by Pearl. The main goal of A* is to find a cheap cost graph path between two nodes in a graph using a heuristic function. The main goal of the heuristic function is to minimize the selection list at each step according to a logical and applicable criterion. In the graph example, finding the shortest path from one node to another has to be done by getting all possible paths and choosing the best. This process is very expensive and time consuming in the presence of a huge number of nodes. On the other hand, using an evaluation function (heuristic function) to minimize our choices according to intelligent and practical criterion would be much faster especially for applications requiring quick output as the driving direction application.

The heuristic function is not a constant static function. It is defined according to the problem in hand and passed to the A* as a parameter.

In the case of A* search for a direction path, the heuristic function F is built up from two main factors:

H = Straight Line distance to destination (distance between two coordinates). G = Distance Traveled so far.

F = H + G.

At each node n, we compute F (n) is computed and the next step is chosen accordingly (the node with the least value of F is chosen).

2.3.2. The *A* algorithm*

A*(Origin, Destination, F)

1. Define a List L that includes all visited nodes ni with their values of F(ni)
2. Define the Stack S that includes nodes ni with their values F(ni)
3. Start at origin (origin is the starting point)
4. Mark the origin as visited
5. Push origin in the stack S
6. Add origin and F(origin) to L
7. Get top element TE of the Stack S
8. For each unmarked neighbor UNi of TE add UNi and F(UNi) to L
9. From L choose N: the node with the least F(N) then pop all elements in S until predecessor of N appears on the top
10. Push N in the stack S

11. Go back to step 5 until the destination node appears or no more unmarked nodes exist

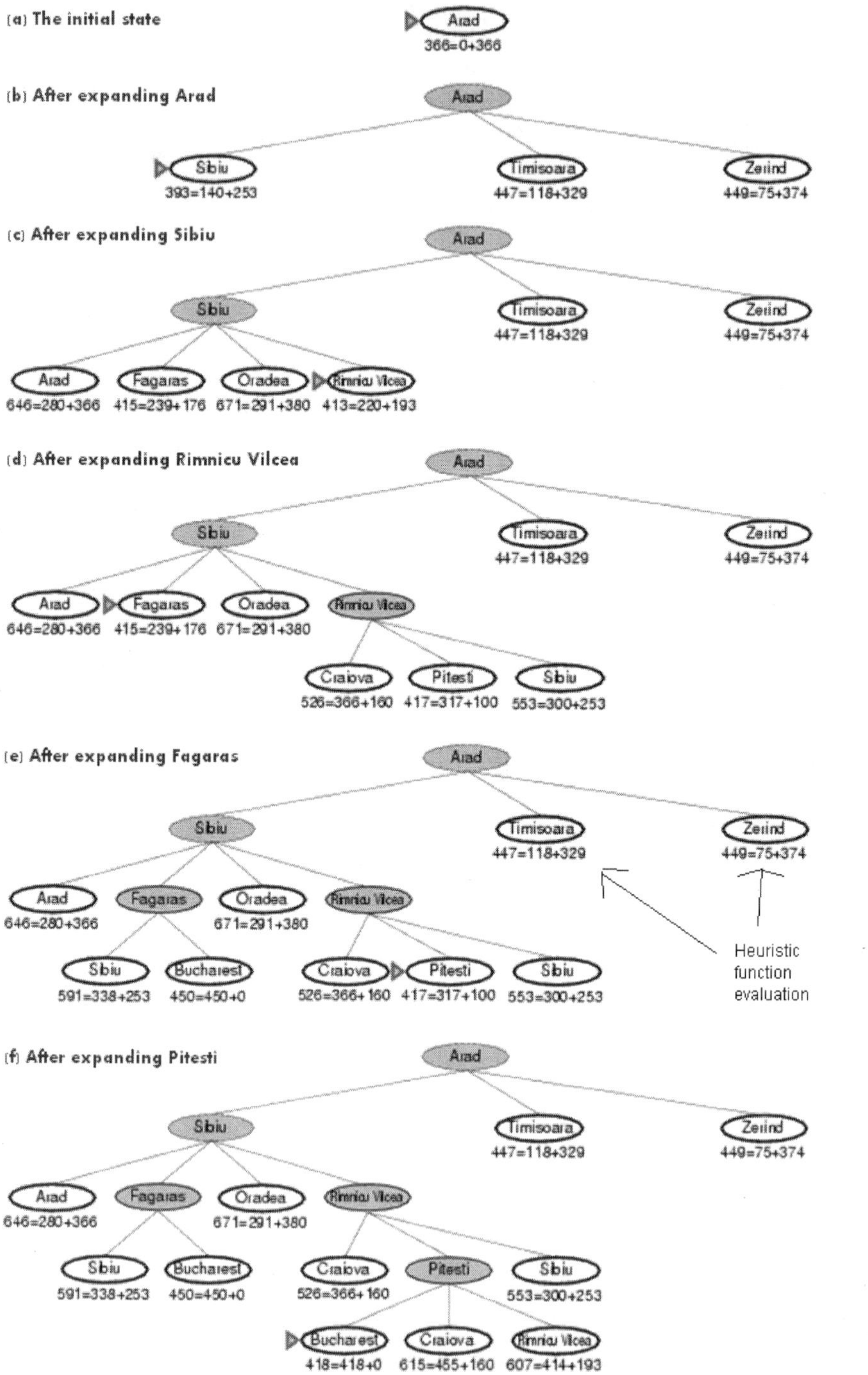

Figure 2. A* algorithm behaviour to find a path starting from "Arad" to "Bucharest"

12. If no more unmarked nodes exist, return "No Solution" otherwise return the Stack content as a solution

Note that A* Algorithm is a polynomial time algorithm with time complexity in O(n2) in the worst case and O(n*logn) in the average and best cases.

Figure 2 is an example of the A* algorithm behavior to find a path starting from "Arad" to "Bucharest" in Romania [4]. First, we start at Arad and go to the next neighbor with the best heuristic function (Sibiu). Second, explore all neighbors of Sibiu for the best heuristic function (evaluation of the function is shown). The algorithm continues to choose the best next step (with the least value of heuristic function) until it reaches Bucharest.

2.3.3. A*traffic: A variation of A* with road traffic as a factor

A*Traffic could be seen as a variation of A* with the ability to take traffic into consideration when computing the driving direction solution. The main job is done in the heuristic function where a new factor is used to choose the next step. The new factor is the average traffic value (got online from real time databases) represented in the following form time/ distance (exam- ple: 3 min/km).

The new Heuristic function will be:

F = H + G + T

Where:

H = Straight Line distance to destination (distance between two coordinates). G = Distance Traveled so far.

T= Average Traffic delay

2.3.4. Testing tool: Query example

This sub-section presents the layout of the testing tool developed to test the algorithm proposed "A*Traffic". For this purpose, an example query is presented.

2.3.5. Query example: From "HU, Kantari St, Hamra" to "AUB, Bliss St, Hamra" (Beirut, Lebanon)

This example demonstrates the main feature of the software. It provides driving directions between "HU, Kantari St, Hamra" (Haigazian University, Beirut, Lebanon) and "AUB, Bliss St, Hamra" (American University in Beirut) in Beirut, Lebanon. In order to find the driving directions, the user has to provide a start address and a destination address and clicks on the "Go!" button. Once the button is clicked, the software generates a path (in blue) between the start and the destination addresses. The blue path generated is a short path (using A*Traffic) to follow in order to drive from the start address to the destination address. Figure 3 illustrates this example.

3. Road networks with time-weighted graphs

This section includes our approach in presenting the intelligent traffic system. Our main idea is to construct a time graph. We mean by the time graph: a graph representing the map with edges weighted by numbers (minutes) representing the estimated time needed to drive the edge (represent a road or part of it). The section starts by presenting the

"Time-weighted Graph", shows a possible example of the graph, gives an execution example when applying the dynamic A*Traffic algorithm proposed in our previous work.

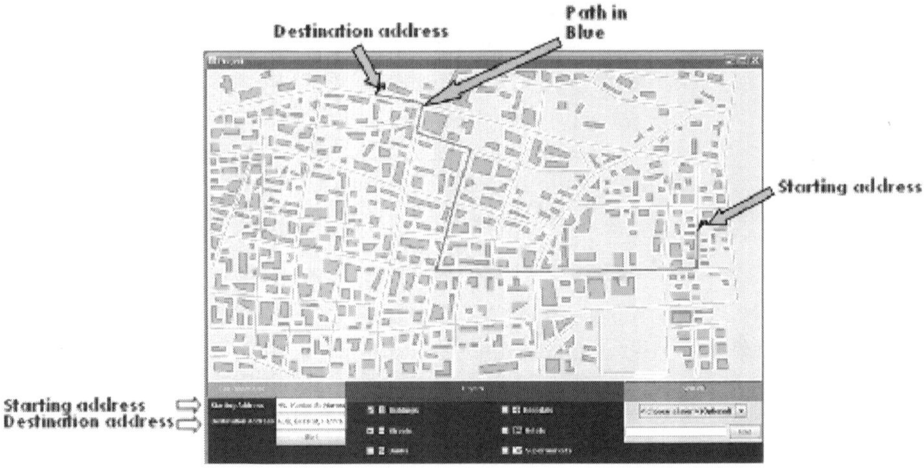

Figure 3. Path from "HU, Kantari St, Hamra" to "AUB, Bliss St, Hamra"

3.1. Time-weighted graph

This section includes our graph proposal using time-weights computations possible examples and executions.

3.1.1. Time weights

Graphs are usually weighted with distances. In this chapter, time will be used as the weight of the graph edges. The main issue is: how to compute the graph edge weight in terms of time (minutes)?

To answer this question, we make the following assumptions:

- Each edge in the graph represents a road, street, highway, etc. or part of it
- Each of these (road, streets,..) has a maximum speed limit that is stored in the database As a result:

The initial weight of the edge (minutes) = (edge distance (miles) ÷ speed (miles/hour)) * 60

3.1.2. Example: Part of Manhattan in a time-weighted graph

To clearly present the suggested idea, an image demonstration will be presented to show how a graph is built and weights are assigned. The following series of images show some part of Manhattan (New York, USA) in map, the process of creating vertices and edges, assigning weights for edges (using stored data) and finally applying the dynamic A* algorithm on such example. Figure 4 shows the map of Manhattan (from Google maps) and the part where the test example was done.

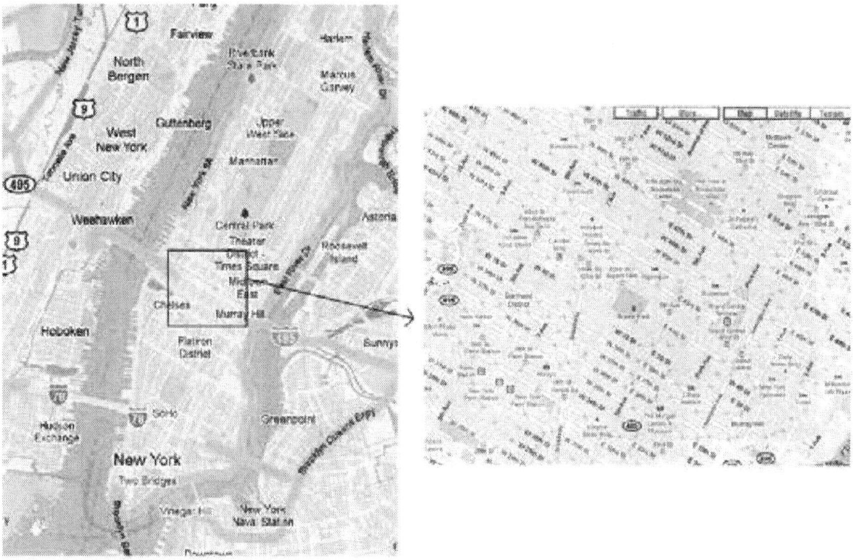

Figure 4. Manhattan and the chosen part (to be modeled in a graph)

The following figures (5, 6, and 7) show the following:

- ❖ Location of vertices in the map (for simplicity, only intersection were chosen)
- ❖ Vertices only
- ❖ And finally the whole graph: vertices and edges

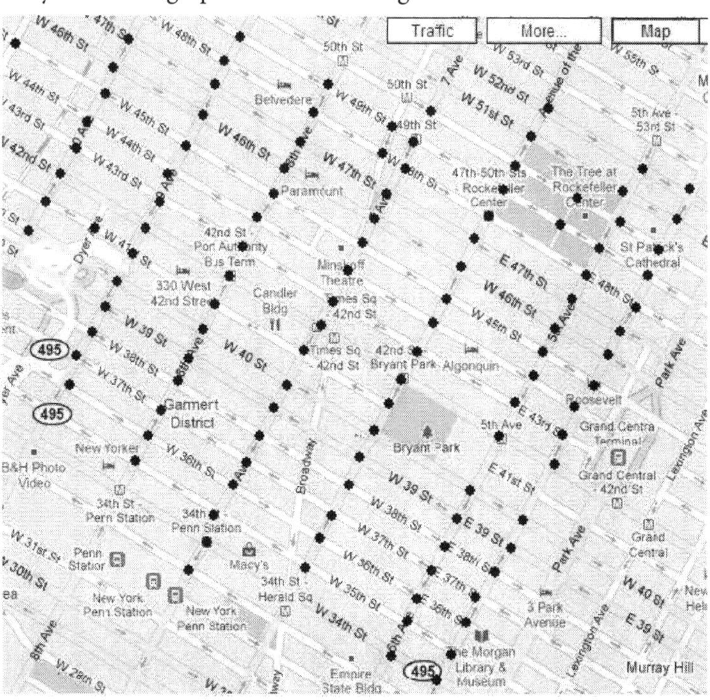

Figure 5. Modeling graph vertices in the chosen part

Figure 6. Considering graph vertices only

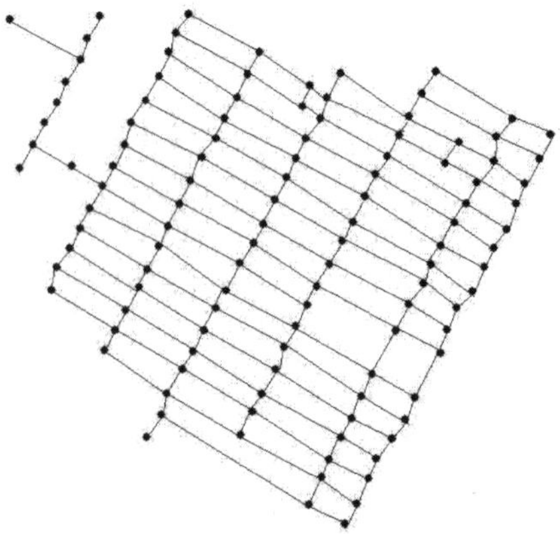

Figure 7. Building the graph edges (directed)

Figure 8 shows one of the edges with weight. The weight is computed as described in section

3.1.1. As a result the shown weight was computed as follows: w = 0.22 (distance) * 30 (speed) / 60 = 0.11 minutes

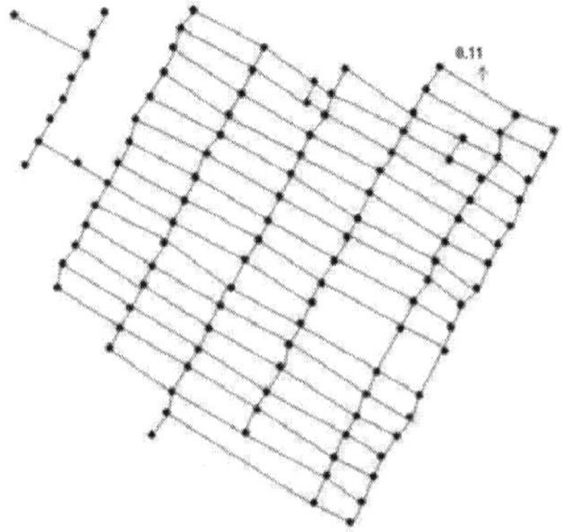

Figure 8. A sample edge with a calculated weight

3.1.3. Online updates with dynamic A*traffic

Dynamic A*Traffic assumes receiving online data whenever a related change occurs. In our new graph, the traffic system assumes receiving online data about current road situations in terms of time units. The main question is:

How can we describe the road traffic changes in time units (minutes)?

If the road is categorized as "heavy traffic", the online system simply does the following:

i. Calculates the average speed (AV) of moving vehicles (not exceeding the max limit) in the heavy traffic road
ii. Get the distance (D) of the road (or the part) where heavy traffic exists
iii. Sends T where T = (AV/D) * 60

3.1.4. Query Example

Figure 9 (a) shows a calculated query path (using A* algorithm) where figure 9 (b) shows a recalculation of the same query (with a new path) after online information about the traffic situation is received. Here are some hints of the shown calculations:

❖ The calculated time T1 in figure (a) is 3.35 minutes
❖ After receiving updated data, T1 became 5.30 minutes
❖ Recalculation is done and another path (figure (b))with 3.83 minutes was found

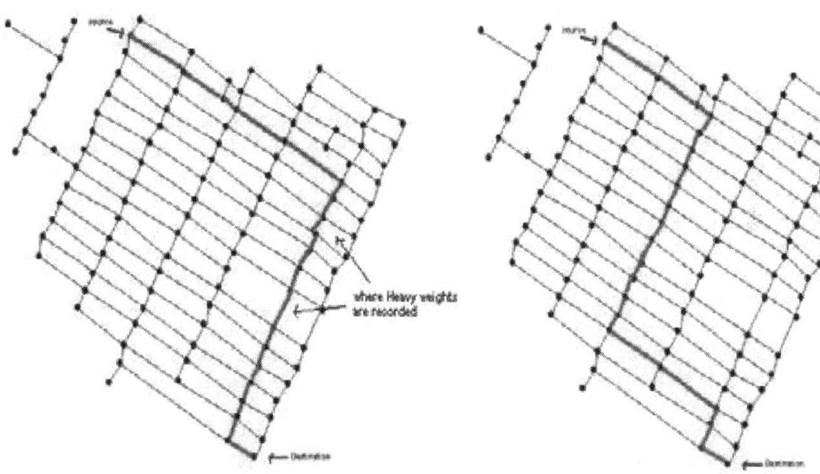

(a) (b)

Figure 9. a) A short (time) path using A*.(b) Another path for the same query after heavy traffic is recodedAnalysis, Results, and Conclusions

In brief, using dynamic A*Traffic algorithm and applying it on time-weighted graphs has advantages such as:

❖ Saves a lot of execution time when finding the path. In an optimal algorithm all possible paths have to be found and the shortest is selected among them. Such an optimal algorithm is not in P (class of polynomial time algorithms) which could take years to solve in some cases. In our case, A* and A* traffic are in P. They guarantee finding a good solution but do not guarantee an optimal solution.

- ❖ Applying time-weighted graphs takes into consideration distance and speed and as a result the algorithm will return the fastest path rather than the shortest path
- ❖ Similar to A*Traffic, our analysis showed that our solution is optimal in 88% of the times and 97% for short driving paths. The reason for such good results is that the A* algorithm takes a lot of path related issues into consideration.

Note that our analysis now is focused on timing rather than distance and hence an optimal solution is a solution with minimum time rather than minimum distance.

Table 1 represents the results gathered from applying 100 executions in each case (long, average, and short) where:

- ❖ Optimal solution: Best solution
- ❖ Good solution: takes maximum of 30% more time than optimal solution
- ❖ Bad solution: Takes more than 30% more time than optimal solution

Table 1. Percentages of quality of solutions over different cas References

Distances	Optimal solution	Good Solution	Bad Solution
Long distances (>300km)	76.4 %	14.2%	9.4%
Average Distances (between 100km and 300km)	90.5%	7.2%	2.3%
Short Distances (<100km)	97.2%	2%	<1%
Average	88%		

References

1. Hatem Halaoui Spatial and Spatio-Temporal Databases Modeling: *Approaches for Modeling and Indexing Spatial and Spatio-Temporal Databases.* VDM Verlag, (2009).
2. Pearl, J. (1984). Heuristics: Intelligent Search Strategies for computer Problem Solving. Addison Wesley, Reading, Massachusetts.
3. Shekhar, S, & Chawla, S. Spatial Databases: A Tour. Prentice Hall. Upper Saddle River, NJ, (2003).
4. Stuart Russell and Peter Norvig *Artificial Intelligence a Modern Approach.* Prentice Hall, Upper Saddle River, New Jersey, (2003).
5. Hatem Halaoui(2008). *Towards Google Earth: A History of Earth Geography".* Book chap- ter (Chapter XVI), Information Systems Research Methods, Epistemology, and Appli- cations, IGI Global.
6. Halaoui Halaoui. Spatio-Temporal Data Model: Data Model and Temporal Index Us- ing Versions for Spatio-Temporal Databases. Proceedings of the GIS Planet 2005, Es- toril, Portugal, (2005).
7. Hatem Halaoui AIRSTD: An Approach for Indexing and retrieving Spatio-Temporal Databases". LNCS (IEEE/ACM SITIS 06), Springer-Verlag, (2007).

Index

8. Hatem Halaoui A Spatio-Temporal Indexing Structure for Efficient Retrieval and Manipulation of Discretely Changing Spatial Data". Journal of Spatial Science (2008)., 53(2)

9. Google Earth Explore, Search and Discover.Available: http://earth.google.com/tour/index.html.Access date 22 December (2009).

10. Google Earth Blog Google Earth Data Size Live Local, New languages coming. Available: http://www.gearthblog.com/blog/archives/2006/09/news_round- up_google.html.Access date 15 December (2009).